Ergebnisse der Physiologie · Reviews of Physiology

Ergebnisse der Physiologie

Biologischen Chemie und experimentellen Pharmakologie

Reviews of Physiology

Biochemistry and Experimental Pharmacology

65

With 45 Figures

Springer-Verlag Berlin Heidelberg GmbH 1972

ISBN 978-3-540-05814-4 ISBN 978-3-540-37462-6 (eBook)
DOI 10.1007/978-3-540-37462-6

Originally published by Springer-Verlag Berlin Heidelberg New York in 1972
Softcover reprint of the hardcover 1st edition 1972

Library of Congress Catalog Card Number 62-37142.

Universitätsdruckerei H. Stürtz AG, Würzburg

Contents

List of Contributors

Bouman, M. A., Prof. Dr., State University of Utrecht, Dept. of Medical and Physiological Physics, NL-Utrecht

Crone, Chr., Prof. Dr., Institute of Medical Physiology Dept. A, University of Copenhagen, Juliane Mariesvej 28, DK-2100 Copenhagen

Koenderink, J. J., Prof. Dr., State University of Utrecht, Dept. of Medical and Physiological Physics, NL-Utrecht

Kruhøffer, P., Prof. Dr., Institute of Medical Physiology Dept. B, University of Copenhagen, Juliane Mariesvej 28, DK-2100 Copenhagen

Pitton, J.-S., Dr., Institut Universitaire de Microbiologie Médicale, 22, Quai de l'École-de-Médecine, CH-1211 Genève 4

Stanbury, J. B., Prof. Dr., Unit of Experimental Medicine, Department of Nutrition and Food Science, Massachusetts Institute of Technology, Cambridge, MA. 02139/USA

Einar Lundsgaard, 1899–1968

Poul Kruhøffer and Christian Crone

Einar Lundsgaard, professor of Physiology at the University of Copenhagen, was a leading figure in that dramatic chain of events which constituted what A. V. Hill called "the revolution in muscle physiology". By a major intellectual achievement he changed the concepts of metabolic energy transformation and he will forever remain one of the great figures in the gallery of Physiologists of Metabolism.

Lundsgaard was born in Copenhagen in Denmark just before the turn of the century as the son of a doctor in private practice. Among his ancestors were several men of law, mostly civil servants. His father had been occupied with research and had received the Gold medal of the University of Copenhagen for a scientific study of prostatic hypertrophy. Lundsgaard would later receive a similar distinction at the same university.

His mother, who was of the Salomon family, was a strong and dominating personality who made heavy demands on her children to do their duty and accomplish what was necessary to consolidate the social position. Altogether Lundsgaard grew up in a climate which was rather common among well-to-do middle class academics in the beginning of this century where hard work and concentrated intellectual effort were rated very highly. This atmosphere undoubtedly marked Lundsgaard—as so many others of his generation—and made of him a rather closed and even inaccessible person. In his youth he was a more outwardly directed person than later when a slightly depressive temperament became more evident. This led to a rather sceptical attitude towards the utility of science and also towards the increasing interest of politicians in academic affairs.

He looked upon research as a fundamentally personal effort and probably never felt really at ease in collaborative so-called team work. He had pupils, but never tried to create a "school". His most prominent Danish pupil was Herman Kalckar who in the thirties worked on phosphorylation in Lundsgaard's laboratory. Even Fritz Lipmann calls himself a pupil of Lundsgaard—although they never worked formally together.

Lundsgaard's special abilities as a scientist stemmed from an unusual talent to reduce a problem to its essence. He had an analytical mind with great sense for generalization and a striking flair for logical deductions. All this gave him an unusual power for rigorous formulation of problems and for profound interpretations of experimental observations. These special intellectual capacities flourished in full richness when in the early 1930's he revolutionized the concepts of the chemistry of muscular contraction. Later, his rigorousness reduced his experimental zeal. It was as though he—by his penetrating analyses—hampered his own research by submitting potential experiments to the very hard test it was to pass through his analytical brain. As he once said when he received the great Scandinavian distinction, the Anders Jahre prize, "I have two Natures within myself, Aladdin and Noureddin. Although it is Aladdin who has brought me the most important results, I feel rather closer to Noureddin, who by intellectual and analytical achievement, won his victories."

Already as a medical student LUNDSGAARD began to work in the Institute of Medical Physiology in the University of Copenhagen. This institution, which had been founded by CHRISTIAN BOHR, was at that time in the hands of VALDEMAR HENRIQUES, whose main research fell within the realm of metabolism. LUNDSGAARD was a devoted pupil of HENRIQUES, whose fine character impressed him deeply, and he started research within his teacher's field.

LUNDSGAARD'S first scientific work dealt with the ability of protein and amino acids—administered in various doses and by various routes—to produce a rise in the blood sugar concentration in rabbits, dogs and human beings. The effects of ammonium salts were also studied.

These studies formed a natural starting-point for subsequent work which in 1929 brought him a doctorate of medicine. The subject was the *specific dynamic action of foodstuffs*, in particular that of proteins. The dissertation, written in Danish, gave a critical and very comprehensive survey of the literature and contained data from numerous experiments on the effects of amino acids and a number of nitrogen-free organic substances on the oxygen uptake of various animals. The most prominent observation was that various amino acids whether administered perorally or parenterally produced very nearly the same increment in oxygen uptake per gram of nitrogen administered; moreover, a similar increment was elicited by ammonium lactate, whereas sodium lactate had little effect on the oxygen uptake. Thus the findings clearly indicated that the action of amino acids to accelerate oxygen uptake is associated with the amino groups, not with the carbon skeletons. No definite conclusion was drawn as to the more intimate nature of the phenomenon, but LUNDSGAARD later (1942) investigated the effect of amino acids and ammonium salts on the O_2-consumption of isolated, perfused livers. Finding here increments in O_2-uptake per gram of nitrogen metabolized similar in magnitude to those observed in intact animals per gram of nitrogen administered he concluded that the high specific dynamic action of amino acids is mainly due to energy used in transformation of ammonia into urea.

By good fortune, one may say, LUNDSGAARD'S early investigations on the specific dynamic action became the breeding ground for his most important scientific contributions: the discovery of the *blockade of glycolysis by monoiodoacetate*, and the exploitation of this tool to show that *the energy of muscle contraction is based on phosphate bond energy*.

The good luck came during his studies on the specific dynamic action of various substances. He decided to investigate the specific dynamic action of monoiodoacetate. At that time there was great interest in the metabolic effects of iodinated compounds, because of the discovery of iodine in thyroxin. LUNDSGAARD was interested in glycine substituted with iodide, but could not get hold of the substance, so instead he used what he could get locally, monoiodoacetate. When injected into animals a universal muscle spasm developed ending with death of the animal and complete muscle stiffness (rigor). This observation became the starting point for important discoveries of the energetics of muscular contraction.

In a series of experiments and papers LUNDSGAARD demonstrated that lactic acid formation is not indispensible for muscle contraction and that other processes, including the splitting of creatine phosphate, serve as energy-suppliers more closely related to the contraction process proper.

The importance of LUNDSGAARD'S work for the advancement of muscle energetics can only be fully appreciated with some knowledge of the concepts which prevailed when his first paper appeared in Biochemische Zeitschrift in 1930; at this time he was thirtyone years of age. Some of the main points may be briefly outlined as follows:

Following FLETCHER and HOPKINS' fundamental studies (1907) on the formation and disappearance of lactic acid in muscles the view had become widely accepted that glycolysis constitutes the direct energy-supplying process for muscle contraction and thus is inseparably linked with the mechanical process. At the end of the twenties this view was still maintained by OTTO MEYERHOF and A. V. HILL, who had shared the Nobel Prize in 1923. The nearly constant relation between the mechanical performance and the lactic acid formation in contracting muscles (near constancy of "the isometric coefficient for lactic acid") was one of MEYERHOF'S prominent arguments.

In the latter part of the twenties there had been voices of doubt. In 1925 EMBDEN reported formation of considerable amounts of lactate in muscles *after* the termination of a tetanic stimulation under anaerobic conditions and therefore claimed that muscle contraction and lactate formation were not inseparably linked together. MEYERHOF, who—contrary to EMBDEN—used indirect stimulation, was unable to confirm these findings. He therefore strongly repudiated EMBDEN'S argument claiming that the post-stimulatory formation of lactate had been caused by overstimulation leading to a post-stimulatory contracture. Attempts by EMBDEN and co-workers to design experiments giving more indirect support for their view received strong criticism from MEYERHOF and HILL.

Some important discoveries had been made within the few years preceding the appearance of LUNDSGAARD's paper. Creatine phosphate ("phosphagen") had been independently discovered in muscle by P. and G. P. EGGLETON and by FISKE and SUBBAROW in 1926. The EGGLETON's reported some breakdown of the substance during contraction, and resynthesis during aerobic recovery conditions. Two years later they described the breakdown to be greater, relative to lactate formation, in the earlier than in the later stages of activity. Breakdown during contraction and partial resynthesis during recovery in muscles exposed to anaerobic conditions was reported in 1928 by NACHMANSOHN working in MEYERHOF's laboratory. The heat of hydrolysis of creatine phosphate had been found by MEYERHOF and LOHMANN to be considerable (some 11–12 kcal/mole). In spite of these findings no definite role had been ascribed to creatine phosphate in muscle contraction—and this also applied to ATP (adenylpyrophosphate) discovered independently in muscle in 1929 by LOHMANN and by FISKE and SUBBARROW.

From MEYERHOF's monograph "Die chemischen Vorgänge im Muskel", which appeared 1930, it is evident that at the time of writing he was still convinced of the essential role of glycolysis in muscular contraction. It is equally clear that the recently acquired knowledge of creatine phosphate—including that provided by himself—was causing trouble and was only reconciled with the accepted view by means of some purely hypothetical assumptions. According to one of these creatine phosphate should undergo no actual breakdown during muscular contraction but only a "labilization" not associated with a release of the heat of hydrolysis.

When in his early studies, performed in Copenhagen, LUNDSGAARD subjected the rigor-muscles of monoiodoacetate poisoned animals to a closer examination, he found, by a simple colorimetric method, that they were no more acid than those of normal, resting animals; and that no shift in reaction towards the acid side occurred in them after death. Concordantly, the lactate concentration was found not to differ from that of normal, resting muscles and no post-mortal increase in lactate content could be demonstrated.

Realizing what a powerful tool monoiodoacetate might be in a closer inquiry into the chemistry of muscle contraction LUNDSGAARD went on to study in poisoned frogs the performance of muscles in which rigor had not developed; inactivation by denervation prior to the poisoning was found to delay markedly the onset of rigor and proved useful in securing muscles in this condition. It turned out that on electric stimulation they were capable of performing an appreciable number (50–100) of twitches without an increase in lactate content—while a rather large increase occurred in unpoisoned muscles at a similar mechanical performance. This observation made it immediately clear that glycolysis is not essential for muscle contraction.

LUNDSGAARD went further in his first paper by showing that in poisoned muscles performing an exhausting series of contractions a complete breakdown of creatine phosphate took place. Simultaneously there occurred an equivalent increase in other organic phosphates (according to analysis mainly hexose-monophosphates) but no change in inorganic phosphate content. The last observation showed incontestably the untenability of MEYERHOF'S postulate that creatine phosphate undergoes only labilization and no splitting during muscle contraction. Altogether the observations published in his first "muscle paper" formed a sound basis for LUNDSGAARD to advance the hypothesis that a splitting of creatine phosphate is the direct energy-supplying process for muscle contraction and that the process of lactate formation is initiated by this splitting and provides energy for a resynthesis of creatine phosphate. By this hypothesis LUNDSGAARD directed attention to "*energetically coupled reactions*" and to *a phosphorylated compound as an energy source*. These concepts became, as is well-known, immensely fruitful for biochemical research. It is natural to mention in this context that LUNDSGAARD was also the first to herald *oxidative phosphorylations*. This happened later, when, from the observation that the working capacity of monoiodoacetate poisoned muscles was appreciably greater in aerobic than in anaerobic conditions, he drew the conclusions that resynthesis of creatine phosphate can be driven by oxidative processes.

LUNDSGAARD informed MEYERHOF of his findings prior to publication, and in the wording of FRITZ LIPMANN, who was then working with MEYERHOF, the message "shook up the Meyerhof laboratory". It redounds to MEYERHOF'S honour that his reply was an invitation to LUNDSGAARD to carry out further studies in his laboratory. So LUNDSGAARD went to the lion's den, which had just been moved from Berlin to Heidelberg. There he met a modern laboratory accommodating a number of outstanding persons, K. LOHMANN, D. NACHMANSOHN, H. BLASCHKO, DEAN BURK, FRITZ LIPMANN, and SEVERO OCHOA to name some. During his five months' stay he extended his studies by experiments performed on pre-rigor muscles isolated from poisoned frogs and maintained under strictly anaerobic conditions. Besides confirming the absence of lactate formation during contractions, he determined the correlation between the mechanical performance and creatine phosphate breakdown. The results showed that, contrary to recent findings of others in unpoisoned muscles, the breakdown of creatine phosphate ran fairly parallel to the contraction performance; yet there was some tendency towards a decline in the breakdown towards the end of an exhausting series of contractions. Despite this small deviation the data were taken by LUNDSGAARD as further evidence for his hypothesis that creatine phosphate breakdown constitutes the direct energy-supplying process for muscle contraction.

With his own contribution to the collapse of another hypothesis in mind he was, however, very careful to emphasize that his, too, might be just another approximation towards the truth. In fact, he pointed out as a possibility that both lactic acid formation and creatine phosphate breakdown might be coupled with a third process, which in turn was directly coupled with the contraction process proper. Actually he was quite close to being the first suggesting ATP breakdown-resynthesis to be the third component of the energy delivering machinery. He had also found out with the imperfect methods then available that, besides a tendency towards diminishing breakdown of creatine phosphate with approaching exhaustion, some breakdown of ATP occurred towards the end of an exhausting series of contractions.

During some later studies he also got near to disclosing the position of ATP. In his 1934 paper (in the Festschrift to HENRIQUES) the breakdown of ATP towards the end of an exhausting series of contractions was found to amount to about half of the initial content. A diminished creatine phosphate breakdown (per unit of mechanical performance) was also found for the pre-exhaustion period and the calculated fall in energy (heat) release from this process was rather closely matched by the calculated energy (heat) release from the ATP breakdown. Yet, because determinations of the fate of ATP in experiments of different design did not fit into the picture, he desisted from proposing ATP an immediate energy source for the contraction mechanism proper. This proposal was made by LOHMANN in the same year on the basis of his finding that ATP or AMP is required for enzymatic breakdown of creatine phosphate in muscles extracts.

In the 1934 paper it was also reported that in poisoned muscles kept anaerobic at low temperature a considerable creatine phosphate breakdown takes place *following* a short tetanus. Furthermore, the ratio between the creatine phosphate breakdown observed *during* the tetanus and the heat liberation calculated from the mechanical performance was found to be considerably smaller at low temperature than at room temperature. In other words, during the tetanus there arose a "debt of creatine phosphate breakdown" which was "paid" during the recovery. By these findings LUNDSGAARD became the first to provide clear evidence against his own working hypothesis of creatine phosphate as the immediate energy source for the contraction process and he concluded that creatine phosphate breakdown, like glycolysis, is only a restitution process.

In earlier studies, important results were also obtained on unpoisoned muscles under anaerobic conditions. It was found that during a series of isometric contractions a gradual rise in lactate formation "goes hand in hand" with the gradual fall in creatine phosphate breakdown previously reported by NACHMANSOHN. Moreover, the isometric coefficient of heat calculated from the heat which should be formed by the two processes together was found to be almost constant throughout the contraction series and quite close to that determined directly by A. V. HILL in myothermic experiments. These findings gave strong support to the view that glycolysis and creatine phosphate breakdown are the only quantitatively important energy-supplying processes during anaerobic conditions, and that the former process is triggered by the latter. Other experiments revealed that during recovery from a short tetanus not only is creatine phosphate resynthesized but lactate is formed as well. Calculations from the available thermochemical data showed that the heat absorption by the former process should be very nearly identical with the heat release by the latter. LUNDSGAARD'S interpretation that the efficiency of the coupling between the two processes is close to unity is subject to criticism. His finding was, however, important because it showed that the already known delayed anaerobic resynthesis of creatine phosphate, rather than being inconsistent with the existence of a small but positive "anaerobic delayed heat", is a component of a reasonable explanation of that phenomenon. Together with other of his findings it served to bring the interpretation of myothermic data out of a deadlock.

LUNDSGAARD essentially stopped his experimental work in muscle physiology in 1934. Motivated, undoubtedly, by his experience with monoiodoacetate he went on to study the metabolic effects of phlorizin with the intention of elucidating the *mechanism of active transepithelial transport of glucose* (hexoses) in the intestine and the renal tubulus. At that time VERZAR'S "Umbauhypothese" had just been given a specific formulation by WILBRANDT and LASZT: that active intestinal absorption of hexoses is brought about by phosphorylation and dephosphorylation in the epithelium. Furthermore POULSON had attributed phlorizin-induced glycosuria to inhibition of the tubular reabsorption, and KAYS had shown that the kidney and the intestinal mucosa have a high content of phosphatase in common. LUNDSGAARD showed that the intestinal absorption of glucose—but not of amino acids—can be markedly inhibited by introduction of phlorizin in the intestine. He was also able to show that in different tissues phlorizin in similar concentrations exerts a profound inhibition of various phosphorylations and dephosphorylations associated with the intermediary carbohydrate metabolism. This led him to accept WILBRANDT and LASZT'S hypothesis and to propose that phlorizin blocks the intestinal absorption of glucose by inhibiting its phosphorylation or the dephosphorylation of glucose phosphate. On the basis of similar evidence this view was also extended to the tubular reabsorption of glucose. The phosphorylation-dephosphorylation hypothesis and LUNDSGAARD'S view on the mechanism of action of phlorizin on the transepithelial transfer of glucose were later shown to be untenable. For many years they served, however, as potent stimuli to research, which in LUNDSGAARD'S laboratory led to valuable informations on the formation of phosphate esters in the intestinal mucosa.

From LUNDSGAARD'S phlorizin papers one point deserves special mention. In muscle extracts phlorizin was found to reduce the formation of lactate from glucose considerable more than the lactate formation from glycogen. From this LUNDSGAARD concluded that the biological breakdown of glycogen does not start with hydrolysis, and, knowing that hexosephosphates can be formed from glycogen, he suggested that the initial process is a *phosphorolysis.*

LUNDSGAARD'S interest in phlorizin led him to study the effects of this substance in eviscerated animals and isolated perfused kidneys, livers and hind limbs, and from the middle of the thirties such preparations became major tools in his own and his associates' investigations of metabolic problems.

The problems to the elucidation of which these tools were first applied were *the effects of insulin and its mode of action.* Before LUNDSGAARD became head of the Institute, NIELS AA. NIELSEN, had worked out there a technique for organ perfusion. In his thesis he had shown that insulin promotes the cellular uptake (utilization) of glucose in perfused hind limbs but had been unable to detect any effect of insulin on the uptake or release of glucose in isolated, perfused livers. These findings were confirmed in greatly extended studies by LUNDSGAARD, NIELSEN and ØRSKOV. In the liver perfusion studies an interesting observation was reported: Whereas one preparation of crystalline insulin was without effect on the course of the glucose concentration of the perfusion blood another crystalline preparation elicited—when the same number of units were added—a prompt and pronounced release of glucose from the liver. This last observation must be regarded as a forerunner of the discovery of glucagon. The results obtained with the perfused livers made LUNDSGAARD spokesman of the view that insulin is without *immediate* effect on the carbohydrate metabolism of the liver. This view was challenged by a Belgian group who claimed that insulin elicits an uptake of glucose in the liver which greatly exceeds that induced in extra-hepatic tissues. Their argument was that maintenance of a "normal" blood glucose concentration in dogs receiving super-maximal doses of insulin had been found to demand infusions of glucose at much higher rates in intact animals than in hepatectomized (or eviscerated) ones. Although LUNDSGAARD realized that it is questionable whether information on possible *direct* effects of insulin on the liver can be obtained by this approach he decided to repeat the Belgian studies using, however, cats as experimental animals. Finding no difference between glucose uptakes of intact versus hepatectomized or eviscerated animals LUNDSGAARD felt further assured that (in cats at least) insulin does not exert any direct effect on the glucose exchange between blood and liver. This question must still be regarded as open.

In the 1930's LUNDSGAARD carried out experiments on eviscerated cats to study the effect of the blood glucose concentration and of insulin on glucose uptake of muscles. He carried out determinations of the glucose concentration in muscles and found that even during maximum rates of glucose uptake the calculated intracellular glucose concentration remained unchanged and close to zero. Since, consequently, there was no parallelism between the rate of cellular glucose uptake and the glucose concentration difference between blood and muscle cells he concluded (1939) that glucose does not pass into muscle cells by simple diffusion and that the uptake-promoting effect of insulin must be exerted on some special ("active") transport process in the cell-membrane. Some years later LUNDSGAARD and coworkers produced further evidence on the nature of the action of insulin and its specificity by showing that in perfused hind limbs, insulin is completely without effect on the cellular uptake of another hexose, fructose.

With his article in 1939 LUNDSGAARD was the first to forward the hypothesis that the main action of insulin on muscle tissue lay in the cell membrane itself. This view gained considerable support in the 1950's through the work of LEVINE and co-workers and is still considered of central importance in theories of the mode of action of insulin.

In 1938 LUNDSGAARD published a short paper which contributed greatly to the knowledge of the *metabolism of alcohol* in the mammalian organism. By liver perfusions it was demonstrated that this organ has a great capacity for oxidation of alcohol to acetic acid. Together with the observation that the elimination of alcohol from the body is almost completely abolished by evisceration this led to the conclusion that the breakdown of alcohol is almost exclusively dependent upon the liver and that the first—and rate-limiting—step is a partial oxidation to acetic acid which is subsequently oxidized completely in extrahepatic tissues. These findings provided a reasonable explanation of the fact that the rate of alcohol elimination is unaffected by muscular exercise. It was later shown by E. J. HOLST—working in LUNDSGAARD'S laboratory—that similar features apply to the metabolism of glycerol, and it was demonstrated by LUNDSGAARD that ethanol suppresses the metabolism of glycerol considerably.

Another of LUNDSGAARD'S associates, N. BLIXENCRONE-MØLLER, studied the *formation of ketone bodies* in perfused livers and showed conclusively, by comparison of the oxygen consumption and the ketone body output, that KNOOP'S theory of formation of only one ketone body per fatty acid was erroneous. The formation of a greater number was another example of a very incomplete oxidation and LUNDSGAARD advanced the view that—in analogy with alcohol and glycerol—the oxidation of fatty acids in the body starts almost exclusively by way of a primary oxidation to ketone bodies in the liver. New information, provided also by a co-worker in his laboratory, made him soon give up this idea, but his interest in the hepatic formation of ketone bodies and in the interaction between the metabolism of the liver and that of the "peripheral" tissues served as an important stimulus to research and to the formation of new concepts.

In 1950 LUNDSGAARD presented evidence that substances formed *extrahepatically affect the metabolism of the liver* and increase its rate of oxygen consumption. During hundreds of liver perfusions he had been intrigued by the fact that during the first half hour of a perfusion the rate of oxygen consumption fell to about half of the initial magnitude. The observation was now made that the initial rate could be re-established by introducing a hind limb preparation into the blood perfusion circuit or by letting a hepatectomized animal serve as blood donor for the liver. LUNDSGAARD was able to show that none of a large number of constituents of blood could be responsible for the phenomenon, but he did not succeed in disclosing its nature. The problem was taken up in the laboratory by IRVING B. FRITZ and although he did not solve in either, it carried him into important research on the metabolic function of carnitine.

As it is seen from this account of LUNDSGAARD'S scientific contribution, he remained faithful to the two tissues whose metabolic processes interested him particularly: *liver* and *muscle* and he made several far reaching discoveries which have greatly helped the understanding of metabolic transformations.

LUNDSGAARD was professor of Physiology at the University of Copenhagen from 1934 until he retired in 1967. He created an institution in which free research was a prominent feature. He never forced his assistants into research projects of his own unless they had a desire to work in his particular branch of the discipline. This attitude had as an effect that when a period of great expansion for Danish Physiology started in the 1960's a group of physiologists, trained in various fields, were ready to lead development into broader areas.

LUNDSGAARD had finished his medical studies in 1923. He published in 1926, at the age of twentyseven, a complete textbook of Physiology, comprising 700 pages. This textbook demonstrated its author's masterful command of his discipline. It was published over the years in 8 revised editions and served as the backbone of physiological knowledge for a whole generation of Scandinavian doctors.

Thanks to his intellectual integrity and his impeccable honesty LUNDSGAARD acquired a strong position in the medical faculty and at an early age he became member of the supreme governing body of the University.

He received international recognition as reflected in various honours bestowed upon him. In 1937 he delivered the Harvey Lecture in New York and became an honorary member of the Harvey Society. He received honorary degrees in the Universities of Montreal (McGill) and Paris. He was president of the 20. International Congress of Physiology in Copenhagen (1950).

In Scandinavia he received many scientific distinctions. As earlier mentioned he received the Anders Jahre prize (1964) in Oslo. In 1960 he received the Thunberg medal in Sweden. He was a member of the Danish Academy of Sciences and Letters since 1938 and became president of the section of Natural Sciences within the Academy in 1962. He served in this position until his death.

His connections with foreign countries were founded when in his youth he spent time in various European laboratories. In 1930 he worked with MEYERHOF, visited EMBDEN and later worked in the Physiological Laboratory at University College, London. He worked for shorter periods of time at the Marine Biological Station in Plymouth and in the Physiological Laboratory in Cambridge.

At his retirement a special Symposium was held in Copenhagen to commemorate his scientific contributions. At this occasion a series of lectures were given by DOROTHY NEEDHAM, DOUGLAS WILKIE, HERMAN KALCKAR, RAGNAR GRANIT, IRVING B. FRITZ, and FRITZ LIPMANN.

Already at the time of retirement LUNDSGAARD was not feeling well and although he continued to work in an office in his old laboratory it was obvious that he was no longer at full force. He probably knew that he was seriously ill, but did not want to admit it, until finally he developed signs of intestinal obstruction. Within a week after admission to hospital he died from an extensive renal carcinoma, in December 1968.

For those who have known LUNDSGAARD he stands as a strong person and an honest scientist who put his stamp on Danish and on International Physiology.

References

1. LUNDSGAARD, E.: En betydningsfuld fejlkilde ved FOLIN og SCHÄFFER's metode til kvantitativ urinsyrebestemmelse. (An important source of error in quantitative determination of uric acid according to FOLIN and SCHÄFFER.) Ugeskr. Læg. **51**, 937 (1923).
2. LUNDSGAARD, E.: Om årsagen til næringsstoffernes specifike dynamiske virkning. (On the cause of the specific dynamic action of food-stuffs.) Thesis (1929).
3. LUNDSGAARD, E.: Researches on the specific dynamic action of proteins. Abstract. 2den Nord. Kongr. f. Physiol. Skand. Arch. Physiol. **55** (1929).
4. LUNDSGAARD, E.: Inwiefern beeinflussen Eiweißstoffe und deren normale Abbauprodukte den Zuckergehalt des Blutes? (How do proteins and amino acids affect blood sugar?) Biochem. Z. **217**, 125 (1930).
5. LUNDSGAARD, E.: Über die Ursache der Aminosäurehyperglykämie. (The cause of hyperglycaemia due to amino acids.) Biochem. Z. **217**, 147 (1930).
6. LUNDSGAARD, E.: Untersuchungen über Muskelkontraktionen ohne Milchsäurebildung. (Muscle contraction without formation of lactic acid.) Biochem. Z. **217**, 162 (1930).
7. LUNDSGAARD, E.: Die Monojodessigsäurewirkung auf die enzymatische Kohlenhydratspaltung. (The effect of monoiodoacetate on enzymatic carbohydrate breakdown.) Biochem. Z. **220**, 1 (1930).
8. LUNDSGAARD, E.: Über die Einwirkung der Monojodessigsäure auf den Spaltungs- und Oxydationsstoffwechsel. (The effect of monoiodoacetate on fermentation and oxidation.) Biochem. Z. **220**, 8 (1930).

9. LUNDSGAARD, E.: Weitere Untersuchungen über Muskelkontraktionen ohne Milchsäurebildung. (Further investigation on muscle contraction without lactate formation.) Biochem. Z. **227**, 51 (1930).
10. LUNDSGAARD, E.: Über die Ursachen der spezifischen dynamischen Wirkung der Nahrungsstoffe. (On the cause of the specific dynamic action of foodstuffs.) I. Skand. Arch. Physiol. **62**, 223 (1931).
11. LUNDSGAARD, E.: Über die Ursachen der spezifischen dynamischen Wirkung der Nahrungsstoffe. (On the cause of the specific dynamic action of foodstuffs.) II. Skand. Arch. Physiol. **62**, 243 (1931).
12. LUNDSGAARD, E.: Über die Bedeutung der Argininphosphorsäure für den Tätigkeitsstoffwechsel der Crustaceenmuskeln. (The significance of arginine phosphate for the activity of Crustacean muscles.) Biochem. Z. **230**, 10 (1931).
13. LUNDSGAARD, E.: Über die Energetik der anaeroben Muskelkontraktion. (The energetics of the anaerobic muscle contraction.) Biochem. Z. **233**, 322 (1931).
14. HENRIQUES, V., LUNDSGAARD, E.: Die milchsäurefreie („alactacide") Muskelkontraktion. (The "alactacide" muscle contraction.) Biochem. Z. **236**, 219 (1931).
15. MEYERHOF, O., LUNDSGAARD, E., BLASCHKO, H.: Über die Energetik der Muskelkontraktion bei aufgehobener Milchsäurebildung. (The energetics of muscle contraction with arrested lactate formation.) Biochem. Z. **236**, 326 (1931).
16. LUNDSGAARD, E.: Betydningen af fænomenet „mælkesyrefrie muskelkontraktioner" for opfattelsen af muskelkontraktionens kemi. (The significance of "alactacid" muscle contractions for an interpretation of the chemistry of muscle contraction.) (Complete translation into English in: HERMAN KALCKAR, Biological phosphorylations, p. 344–353. New Jersey: Prentice-Hall Inc. 1969.) Hospitalstidende **75**, 84 (1932).
17. BAYLISS, L. E., LUNDSGAARD, E.: The action of cyanide on the isolated mammalian kidney. J. Physiol. (Lond.) **74**, 279 (1932).
18. LUNDSGAARD, E.: Weitere Untersuchungen über die Einwirkung der Halogenessigsäuren auf den Spaltungs- und Oxydationsstoffwechsel. (Further investigations of the effect of halogen-substituted acetate on fermentative and oxidative metabolism.) Biochem. Z. **250**, 61 (1932).
19. LUNDSGAARD, E.: Die Glykolyse. (Glycolysis.) Ergebn. Enzymforsch. **2**, 179 (1933).
20. LUNDSGAARD, E.: Nobelprisen i fysiologi og medicin 1932. (The Nobel prize in physiology and medicine.) Naturens Verden (1933).
21. CATTELL, MCK., LUNDSGAARD, E.: The efficiency of isolated muscle in relation to the degree of anaerobic activity. J. Physiol. (Lond.) **78**, 246 (1933).
22. LUNDSGAARD, E.: Hemmung von Esterifizierungsvorgängen als Ursache der Phlorrhizinwirkung. (Phlorrizin as inhibitor of esterification.) Biochem. Z. **264**, 209 (1933).
23. LUNDSGAARD, E.: Die Wirkung von Phlorrhizin auf die Glucose-Resorption. (The effect of phlorrizin on glucose absorption.) Biochem. Z. **264**, 221 (1933).
24. LUNDSGAARD, E.: Phosphagen- und Pyrophosphatumsatz in jodessigsäurevergifteten Muskeln. (Metabolic transformation of "phosphagen" and pyrophosphate in iodoacetate poisoned muscle.) Biochem. Z. **269**, 308 (1934).
25. LUNDSGAARD, E., WILSON, A. T.: Muscle phosphorus compounds in adrenal insufficiency. J. Physiol. (Lond.) **80**, 29P (1934).
26. LUNDSGAARD, E.: Die Glykolyse. (Glycolysis.) Angew. Chem. **47**, 495 (1934).
27. LUNDSGAARD, E.: The effect of phlorizin on the isolated kidney and isolated liver. Skand. Arch. Physiol. **72**, 265 (1935).
28. LUNDSGAARD, E., NIELSEN, N. AA., ØRSKOV, S. L.: The carbohydrate metabolism of the isolated cat liver. Skand. Arch. Physiol. **73**, 296 (1936).
29. LUNDSGAARD, E.: VALDEMAR HENRIQUES, 19. April 1864–4. December 1936. (Obituary.) Hospitalstidende **79**, 1317 (1936).
30. LUNDSGAARD, E.: VALDEMAR HENRIQUES, 19. April 1864–4. December 1936. (Obituary.) Skand. Arch. Physiol. **76**, 101 (1937).
31. HEVESY, G., LUNDSGAARD, E.: Lecithinaemia following the administration of fat. Nature (Lond.) **140**, 275 (1937).
32. LUNDSGAARD, E.: Artificial respiration. Månedskr. prak. Lægegerning og soc. med. p. 3 (1937).
33. LUNDSGAARD, E.: Orienterende bemærkninger over kredsløbsforholdene i underextremiteterne. (Aspects of the circulation in the lower extremities.) Hospitalstidende **80**, 2 (1937).

34. Hahn, L. A., Hevesy, G., Lundsgaard, E.: The circulation of phosphorus in the body revealed by application of radioactive phosphorus as indicator. Biochem. J. **31**, 1907 (1937).
35. Lundsgaard, E.: Alcohol oxidation as a function of the liver. C. R. Lab. Carlsberg, Ser. Chim. **22**, 333 (1938).
36. Lundsgaard, E.: The Pasteur-Meyerhof reaction in muscle metabolism. Harvey Lecture, November 18, 1937. Bull. N.Y. Acad. Med., April, p. 163 (1938).
37. Lundsgaard, E.: The biochemistry of muscle. Ann. Rev. Biochem. **7**, 377 (1938).
38. Lundsgaard, E.: The Pasteur-Meyerhof reaction in muscle metabolism. Harvey Lect. **33**, 65 (1937–1938).
39. Lundsgaard, E.: The phosphate exchange between blood and tissue in experiments with artificially perfused livers and hind limb preparations. Skand. Arch. Physiol. **80**, 291 (1938).
40. Lundsgaard, E.: The chemistry of the anaerobic muscular contraction. Bull. Johns Hopk. Hosp. **63**, 1 (1938). Dohme Lecture.
41. Lundsgaard, E.: The metabolism of the anaerobic working muscles. Bull. Johns Hopk. Hosp. **63**, 15 (1938). Dohme Lecture.
42. Lundsgaard, E.: The metabolism of the isolated liver. Bull. Johns Hopk. Hosp. **63**, 90 (1938). Dohme Lecture.
43. Lundsgaard, E., Nielsen, N. Aa., Ørskov, S. L.: On the utilization of glucose and the formation of lactic acid in the isolated hind limb preparation. Skand. Arch. Physiol. **81**, 20 (1939).
44. Lundsgaard, E., Nielsen, N. Aa., Ørskov, S. L.: On the possibility of demonstrating an effect of insulin on isolated mammalian liver. Skand. Arch. Physiol. **81**, 11 (1939).
45. Lundsgaard, E.: On the mode of action of insulin. Upsala Läk.-Fören. Förh., Ny följd. **45**, 143 (1939).
46. Lundsgaard, E.: Die säurelöslichen Phosphatverbindungen in der Darmschleimhaut bei Ruhe und während der Hexoseresorption. (Phosphate compounds in intestinal mucosa.) Hoppe-Seylers Z. physiol. Chem. **261**, 193 (1939).
47. Liljestrand, Aa., Lundsgaard, E.: Utilization of glucose and phosphate metabolism in hind limb preparations of cats poisoned with phlorizin. Skand. Arch. Physiol. **83**, 121 (1940).
48. Lundsgaard, E.: Über die aerobe Glykolyse des Darmes. (On the aerobic glycolysis of the intestine.) J. Physiol. of the U.S.S.R. **29**, 311 (1940).
49. Kjerulf-Jensen, K., Lundsgaard, E.: Quantitative Wertung des Umsatzes der Phosphatester in der Darmschleimhaut von Ratten während der Fructoseresorption. (Formation of phosphate esters in the intestinal mucosa during absorption of fructose.) Hoppe-Seylers Z. physiol. Chem. **266**, 217 (1940).
50. Lundsgaard, E.: Leverens rolle i stofskiftet. (The liver in intermediary metabolism.) Nord. Med. **10**, 1593 (1941).
51. Lundsgaard, E.: The specific dynamic action of amino acids and ammonium salts. Acta physiol. scand. **4**, 330 (1942).
52. Lundsgaard, E.: Anatomi, fysiologi og biokemi. (Anatomy, physiology and biochemistry.) In: Danmarks Kultur, 1940. Printed by Det Danske Forlag 1942.
53. Kjerulf-Jensen, K., Lundsgaard, E.: Phosphate exchange between blood and muscle tissue under the influence of insulin. Acta physiol. scand. **7**, 209 (1944).
54. Gammeltoft, A., Kruhøffer, P., Lundsgaard, E.: Insulin and the assimilation of fructose. Acta physiol. scand. **8**, 162 (1944).
55. Lundsgaard, E.: De kemiske omsætninger i cellerne. (Cell metabolism.) In: Videnskaben i Dag, p. 197 (1945).
56. Lundsgaard, E.: Universitetet og de højere læreanstalter. (The University during the occupation.) In: Danmark under Besættelsen **1**, 529 (1946).
57. Lundsgaard, E., Fredericia, L. S.: Obituary in: Videnskabernes Selskab, November 28, 1947.
58. Lundsgaard, E.: Glycerol oxidation and muscular exercise. Acta physiol. scand. **12**, 27 (1947).
59. Lundsgaard, E., Schou, S. Aa.: Hormoner og hormonpræparater, vitaminer og vitaminpræparater. (Hormone and hormone preparations.) Arch. Pharm. Chem., p. 3 (1949).
60. Lundsgaard, E.: Insulin and muscle. Rap. II Congr. Int. Thérapeut. juin, 119 (1949).

61. LUNDSGAARD, E.: Homoiostase. (Homeostasis.) Med. Forum **10**, 289 (1949).
62. LUNDSGAARD, E.: In memoriam. AUGUST KROGH. Experientia (Basel) **6**, 39 (1950).
63. LUNDSGAARD, E.: The ATP content of resting and active muscle. Proc. roy. Soc. **137**, 73 (1950).
64. LUNDSGAARD, E.: Observations on a factor determining the metabolic rate of the liver. Biochim. biophys. Acta (Amst.) **4**, 322 (1950).
65. LUNDSGAARD, E.: De Anima. Med. Forum **5**, 21–26 (1952).
66. LUNDSGAARD, E.: Forsøg med dyr. (Experiments with animals.) In: Politiken (newspaper), November 19, 1952.
67. LUNDSGAARD, E.: JOHANNES C. BOCK. 2. Oktober 1867–8. Januar 1953. (Obituary.) In: Videnskabernes Selskab, November 20, 1953.
68. LUNDSGAARD, E.: The liver and carbohydrate metabolism. In: Liver injury. Transactions of the 12th Conference, p. 11. New York: Josiah Macy Jr. Foundation 1953.
69. LUNDSGAARD, E.: Musklernes, nervesystemets, indlæringens og åndedrættets fysiologi. (Selected chapters of physiology.) In: Nordisk Lærebog for Talepædagoger, p. 102 (1954).
70. LUNDSGAARD, E.: Insulin and glucose uptake by the liver. Acta physiol. scand. **31**, 215 (1954).
71. LUNDSGAARD, E.: Physiology in Denmark. In: Perspectives in Physiology. Edit.: I. VEITH. Amer. Physiol. Soc., Washington, D.C. 1954.
72. LUNDSGAARD, E.: Leverens rolle for den samlede organismes stofskifte. (The role of the liver in the metabolism of the organism.) Nord. Vet.-Med. **7**, 869 (1955).
73. LUNDSGAARD, E.: Carbogen og kunstig respiration. (Carbogen and artificial respiration.) Ugeskr. Læg. **47**, 317 (1956).
74. LUNDSGAARD, E.: Om døden fra et biologisk synspunkt. (The death—a biologist's viewpoint.) In: Politiken (newspaper), March 8, 1958.
75. LUNDSGAARD, E.: The physiology of alcohol. Atti Soc. lombarda Sci. med.-biol. **14**, 865 (1959).
76. LUNDSGAARD, E.: Mennesket som dyreart (Man as an animal species). Copenhagen: Munksgaard Publ. 1960.
77. LUNDSGAARD, E.: Lærebog i Fysiologi. (Textbook of physiology.) 1st ed. 1926. 8th ed. 1964. 670 pp. Copenhagen: Arnold Busck Publ.

Mechanisms of Bacterial Resistance to Antibiotics*

J.-S. Pitton**

With 19 Figures

Table of Contents

---

* This work was supported by the "Fonds National Suisse de la Recherche Scientifique" (Grant No. 3.27.68).

** Institute of Medical Microbiology, University of Geneva, Geneva, Switzerland.

I. Introduction

Bacterial resistance to antibiotics and chemotherapeutic drugs is important for two main reasons: first, this phenomenon raises serious problems in the treatment of some infectious diseases; secondly, it forms one of the fundamental aspects in modern microbiological research.

Original studies on drug resistance were carried out by LURIA and DELBRÜCK (1943), DEMEREC (1948), and LEDERBERG and LEDERBERG (1952), who showed that mutations towards resistance can occur spontaneously, without any effect due to the presence of the antibiotic. A drug is a powerful selective agent, promoting survival of resistant cells to the detriment of the sensitive population, after the genetic alteration (i.e. the mutation) and its phenotypic expression have taken place.

It is generally recognized that drug resistance is mediated by biochemical mechanisms. The action of many antibacterial substances has been investigated but the specific cause of drug resistance has been identified in very few cases. It must be stressed that the site of action of an antibiotic is not the same as that of the antibiotic resistance. In recent years, a number of investigations have provided new ideas on these phenomena.

The genetic aspect of the problem has been investigated very thoroughly. The location on the bacterial chromosome of the genes responsible for the different drug resistances has been determined. Moreover, the mechanisms of transfer and acquisition of transferable drug resistance, especially in Gram-negative bacilli, have been described by WATANABE (1963) and ANDERSON (1967). The results have brought to light many interesting new problems. Some workers studied the mode of acquisition and perpetuation of the genetic determinants responsible for drug resistance, while others defined the nature of the resistance, its mode of action at the cellular and molecular level and its relations to the genetic control.

The experiments carried out on the structural components of bacteria (nucleic acids, proteins, polysaccharides, heteropolymers of cell wall) as well as the different enzymatic steps involved in the biosynthetic processes (DNA replication, protein synthesis, structure and function of the ribosomes) have improved our knowledge of the site of action of many antibiotics. Therefore, the biochemical mechanisms involved in drug resistance have been defined more precisely.

The phenotypic modifications responsible for these resistances can result in the following: decreased permeability to the inhibitor, the resistant mutants having lost a system of active transport; increased destruction or inactivation of the drug; formation of an altered receptor in the cell. The other possibilities, increased synthesis of a metabolite antagonizing the drug or increased concentration of the enzymatic system sensitive to the drug, are not accepted mechanisms.

Most resistant bacteria can thus be subdivided into two main groups: *drug-tolerant* cells able to grow in high concentrations of an unmodified antibiotic, and *drug-destroying* cells owing their resistance to the presence of an enzymatic system able to inactivate or even to destroy the molecule of the drug.

II. Acquisition of Drug Resistance

Bacterial cells can acquire drug resistance by different processes.

A. Natural Resistance

Natural resistance limits the anti-microbial spectrum of most antibiotics. Permeability barriers seem likely to offer the best explanation for the definition of the spectrum of a drug. However, the index of permeability, based on the study of gradients in drug uptake, is rather unreliable when the experiments are done with intact cells. A decreased uptake can be due to a reduction in the number of binding sites or a reduced affinity for the antibiotic; no modification of the cell, membrane or cell wall can be postulated.

When an antibacterial agent is used therapeutically, the sensitive clones of any microorganism are suppressed, promoting the spread in the host of naturally resistant strains which initially constituted a small proportion of the population.

Natural resistance is a matter of total or partial tolerance to the antibiotic. This tolerance can be expressed by all the cells of a bacterial species or of a strain, or it can be limited to some strains of the same species or to some cells of the same strain. For example, *Pseudomonas aeruginosa* is resistant to a large number of antibiotics, while the Gram-negative bacilli are resistant to benzylpenicillin but sensitive to other antibiotics.

The mechanisms inducing drug resistance *in vitro* are three: mutation, which is a change in the DNA within the cell; induced phenotypic resistance; and transfer of genetic material, which is a gain of DNA by the cell.

B. Resistance by Mutation

Drug resistance occurs spontaneously, from a single sensitive cell which, by mutation, can acquire the property of growing in drug concentrations higher than those which are normally active on the original population.

Drug-tolerant bacteria normally arise in a discontinuous process, suggesting spontaneous mutation. In the absence of the antibiotic, only some resistant cells can be detected, since the sensitive cells can still grow. In the presence of the drug, which has a selective pressure, a population containing only resistant cells is obtained.

For a long time it was supposed that the emergence of drug resistance in a normally sensitive bacterial culture was due to spontaneous mutation alone. The most efficient methods for demonstrating this hypothesis are due to LEDERBERG and LEDERBERG (1952) with the replica-plating technique, and to LURIA and DELBRÜCK (1943) with their fluctuation test applied by DEMEREC (1948) to the study of drug resistance.

Mutation rate. The mutation rate is defined as the probability per cell division of a given type of mutant appearing. This rate is relatively constant. It is generally accepted that the frequency of mutation towards drug resistance is between $1/10^{10}$ and $1/10^{6}$ bacterial cells. One of the main characteristics of such a mutation is its rarity.

The emergence of resistant mutants can proceed in various steps which define different types of drug resistance; it is usual to distinguish between:

a) Obligatory one-step resistance. In this case, the bacterium is either totally sensitive of totally resistant to the drug.

b) Optional one-step resistance. This is also called streptomycin-like resistance, since the majority of bacterial strains react in this way with this antibiotic or with closely related substances. In this case, some clones in the population show an intermediate resistance level.

c) Step-wise resistance. With the majority of antibiotics, it is quite impossible to obtain high-level resistant mutants spontaneously, but cells showing a step-wise ability to grow in increasing concentrations of the drug can be obtained. These steps are normally two or three times the level of the minimum inhibitory concentration (MIC). This multiple-step process is the most frequent and its danger in therapeutics is a function not only of the frequency of mutation but also of the size of the first step. This step can be below or above the normal pharmacological concentration of the drug in blood or in different organs. This type is also called penicillin-like and is found in all penicillin derivatives, aminoglucosides other than streptomycin, and broad-spectrum antibiotics such as tetracyclines and chloramphenicol.

These various ways of acquiring drug resistance demonstrate the discontinuity of the process, the mutation being located on one gene (streptomycin) or on a polygenic system (penicillin).

The stability of the resistance plays an important role; a resistant clone can normally be subcultured for many generations in a medium deprived of the drug without losing its resistant properties. However, it is incorrect to assume that mutations proceed only from sensitivity towards resistance. A resistant population always contains a certain number of sensitive cells. Nevertheless, it is difficult to determine the probability of the back-mutation since it is often impossible to provide a selective pressure favouring the sensitive revertants. HASHIMOTO and HIROTA (1966) have described a very useful method for *in vitro* selection of sensitive clones in a resistant population.

Some other features can be added to those described above. In bacterial genetics, it is generally agreed that different mutations arising in one cell are specific and independent of one another. The specificity implies that each mutation affectso nly one character, without influencing other characters of the strain. One mutation does not modify the ability (in fact, the probability) of the cell of undergoing a second mutation. This is defined as independence in a bacterial cell. This is important from the statistical point of view, since the probability of a cell undergoing two distinct and simultaneous mutations is equivalent to the product of the individual probabilities for each of these mutations. It will be seen that this fact was important in the development of the hypothesis on transferable drug resistance. However, it must be pointed out that specificity and independence are not always observed in drug resistant mutants. On the one hand, cross-resistances are very often observed, at least within the same groups of antibiotics, and even between them. On the other hand, in many cases some minor modifications follow the acquisition of drug resistance after mutation: modifications of some biochemical or staining characters; decrease in the adsorption of some phages; modification of the growth rate, generally towards a decrease. These phenomena can be of great importance in the ecology of bacterial populations, *in vivo* as well as *in vitro*.

C. Induced Phenotypic Resistance

When an antibiotic is introduced in a growing culture, in a concentration approximate to the MIC, three types of reaction can occur:

1. Growth is resumed by means of a metabolic response within the cell. This type of resistance has been described by MOYED (1960) with 2-thiazole-alanine which exerts a pseudo-feedback inhibition of histidine synthesis but does not repress formation of the histidine biosynthesizing enzymes. The inhibition lowers the intracellular level of histidine and provokes immediate bacteriostasis; this starvation derepresses the enzymes of the histidine pathway, the level of the inhibited enzymes rises and growth is restored.

2. Growth continues because the drug is destroyed or inactivated by the action of an inducible enzyme. The genetic control of biosynthesis of these enzymes will be discussed later. In contrast to genotypic resistance, which normally persists in the absence of the drug, induced phenotypic resistance is lost after some generations in a medium which does not contain the antibiotic, because, lacking any stimulus for production of the enzyme, the excess is diluted out.

3. This reaction represents a real adaptation of the bacteria and is probably the only example of an adaptative process in drug resistance. It is the special case in which penicillins, or other closely related antibiotics, direct the emer-

gence of resistant L-forms by a heritable but non-genotypic modification, since it eliminates one of the cell-wall components which is probably necessary as a "model" for further normal mucopeptide synthesis.

D. Transfer of Genetic Material

In the past ten years, many facts have come to light which are relevant to the transfer of genetic material between bacterial cells. This transfer can be accomplished by three distinct mechanisms:

1. Chromosomal transfer of resistance genes.
2. Transduction of these genes by bacteriophage infection.
3. Transfer of extrachromosomal genes after conjugation promoted by plasmids or episomes.

1. Chromosomal Transfer

This mechanism probably does not play an important role, at least for pathogenic bacteria. Indeed, the only known system in which such a process is effective is thought to occur with streptomycin and its derivatives. In fact, the streptomycin-resistance gene is recessive when present in a heterozygote along with the sensitive allele. The introduction in a sensitive cell of a fragment of chromosome bearing the locus responsible for resistance results in the formation of a merozygote sensitive to the drug. This can be compared with the zygotic induction described by JACOB and WOLLMAN (1956), and WOLLMAN and JACOB (1957) with certain prophages.

2. Transduction

RITZ and BALDWIN (1961) were the first to demonstrate that a phage can transfer the character of penicillinase synthesis. Cotransduction of two resistance determinants often occurs, thus showing the close genetic linkage between the two genes responsible for drug resistance. However, NOVICK and RICHMOND (1965) believe that the transduced DNA fragments are plasmids. Nevertheless, ASHESHOV (1966a) described an experiment with a strain of *S. aureus* in which the genetic material seemed to be chromosomally located.

Transduction mechanisms have also been demonstrated *in vivo* by JAROLMEN et al. (1965) who inoculated mice intravenously with a sensitive strain of *S. aureus*, followed six days later by a transducing phage suspension obtained from a strain resistant to both tetracycline and penicillin. Staphylococci resistant to tetracycline but not to penicillin were isolated from kidneys, their number being increased after tetracycline treatment of the mice. NOVICK and MORSE (1967) inoculated mice with cultures of two strains of *S. aureus;* one lysogenic and possessing erythromycin resistance, the other resistant to streptomycin; they isolated from kidneys bacteria resistant to these antibiotics in much greater numbers when the mice were treated with both drugs.

At present it is not known to what extent the transduction phenomenon is responsible for antibiotic resistance in clinical isolates of staphylococci, but in all probability it plays a major role.

3. Transfer Promoted by Plasmids or Episomes

The most frequent mechanism of acquiring drug resistance in Gram-negative bacteria is probably a cell-to-cell transfer of plasmids or episomes, which are extra-chromosomal elements.

The term extra-chromosomal element must be reserved for hereditary units, physically distinct from the chromosome; however, such an element is a component of the bacterial genome, even if it is not physically linked with the chromosome. The term plasmid will be used as synonymous with an extra-chromosomal element. The word episome seems to be reserved for a certain class of plasmids which can exist either in an integrated state on the chromosome of the host cell, or in an autonomous cytoplasmic state, the F-factor, for example. Certain authors (Harada et al., 1961; Jacob et al., 1965) have used the term plasmid for extra-chromosomal elements incapable of becoming integrated on the chromosome. But, especially with the drug-resistance factors, it is sometimes difficult to demonstrate that such an element is incapable of becoming integrated.

A transferable plasmid is a plasmid bearing the genetic determinants responsible for promoting its transfer by conjugation. A sex factor is a special plasmid able to promote the transfer of genetic units which are not linked to itself. Non-transferable plasmids are elements which are not able to be transferred by themselves, but which can be recombined with an appropriate factor, such as a phage, an F-factor or a transfer factor.

Two types of plasmids can be distinguished. In quiescent plasmids, replication is controlled so that it occurs once per cell division cycle. A vegetative plasmid is one whose replication is not under such control and is restricted only by non-specific metabolic limitations, such as the availability of precursors, energy and replicating enzymes (Novick, 1969), and therefore there can be more than one per cell.

Among all the plasmids known at present-colicinogenic factors, temperate phages, F or F′ factors, determinants responsible for hemolysin or enterotoxin synthesis—we shall dwell on the R-factors responsible for different types of drug resistance in bacteria.

III. The R-Factors

In Japan, the introduction in 1945 of sulphonamides in antibacterial therapy caused a drastic reduction in the incidence of bacillary dysentery. However, after 1949, this trend stopped and the incidence of the disease started to rise again. Moreover, it was noticed that sulphonamide-resistant strains were found more often than sensitive ones. After the introduction of streptomycin, chloramphenicol and tetracycline, the morbidity fell again temporarily. KITAMOTO et al. (1956) pointed out that the strains of *Shigella* involved could exhibit simultaneous resistance to these four drugs. This phenomenon seemed to be responsible for the increased incidence of the disease. Several drug resistances were detected in single bacterial strains and the frequency of isolation of these strains was greater than that possible by mutation, as a strain resistant to four antibiotics simultaneously would have to be the progeny of one out of 10^{28} sensitive cells. It was proposed by AKIBA (1959) that the mechanism responsible was different from mutation. He assumed that a cell could receive all these resistances "en bloc" from other bacteria already present in the gastro-intestinal tract. Soon after, OCHIAI et al. (1959) and AKIBA et al. (1960) demonstrated that this was in fact the case. Thus the existence of R-factors was predicted before their discovery.

MITSUHASHI et al. (1960) and afterwards WATANABE and FUKASAWA (1961 b) showed that the transmission of resistance was mediated by a conjugation process between a multiresistant donor and a sensitive recipient cell.

Soon after, the same properties were observed in different countries: Great Britain (DATTA, 1962), Germany (LEBEK, 1963 b), France (CHABBERT and BAUDENS, 1965), and the United States (KABINS and COHEN, 1966). This phenomenon has now been described all over the world.

It is important to point out that bacterial strains isolated from patients treated with antibiotics in hospitals are more frequently carriers of R-factors than strains isolated elsewhere. The use (and abuse) of antibiotics need not be implicated in the formation of these factors, since some were detected in strains isolated before the commercial introduction of tetracycline, streptomycin or chloramphenicol (ANDERSON, 1967). In spite of this, it is beyond doubt that these drugs played an important role in selection of strains carrying R-factors. For the moment, it seems likely that the presence of transmissible R-factors is limited to Gram-negative bacilli.

These factors can confer resistance to nearly all known antibiotics and many synthetic chemotherapeutic agents used in the treatment of infectious diseases caused by these bacteria. Resistance to kanamycin-neomycin (LEBEK, 1963 a), ampicillin and cephalothin (ANDERSON and DATTA, 1965), furazolidone (SMITH and HALLS, 1966), to the group spectinomycin, gentamycin and viomycin (SMITH, 1967) can be added to resistance to sulphonamides, tetracycline, chloramphenicol and streptomycin. Bacteria which can carry R-factors, called R^+ strains, can be found in all the different enterobacteria or related strains such as *E. coli*, *Salmonella*, *Shigella*, *Proteus*, *Klebsiella*, *Vibrio*, *Pasteurella* and *Serratia*.

A. General Properties of Transferable Drug Resistance

Watanabe (1963) postulated that the genetic determinants responsible for drug resistance are distributed on a linear linkage group, the resistance transfer factor (RTF) being inserted at a certain point in the group. The RTF would be responsible for the transferability of the complex. But, in spite of the linearity, resistance genes could not be separated from one another in interrupted mating mixtures done with these strains. In this system, all resistances were transferred simultaneously or not at all.

It has been confirmed that R-factors consist of two components: the transfer factor (TF) and the resistance determinants (R-determinants). In appropriate conditions, many R-factors can be dissociated during their transfer to sensitive cells, giving normally stable strains, bearing the transfer factor, one (or more) R-determinant, or both (Anderson and Lewis, 1965b; Anderson, 1965a, 1965b, 1968). Anderson proposed that R-factors comprise two independent replicons, the TF and the R-determinant, each able to survive indefinitely in bacterial cells. When a resistance determinant and an adequate transfer factor are present simultaneously in the same cell, association can occur between them, giving rise to a functional, transferable R-factor.

1. Number of R-Factor Copies per Cell

The problem of the number of R-factor copies per cell is not yet resolved. It has been possible to infer from two different types of investigation that in *E. coli* there is only one copy of R-factor per chromosome. Rownd et al. (1966) demonstrated this by physical measurements after separation of DNA fragments on cesium chloride gradients. Anderson et al. (1968) studied the competition between two determinants and concluded that there is probably only one site in the cell which is able to bind with one of these determinants. This implies a very precise location inside the cell and the presence of only one copy of each physiological unit. It is conceivable that, unless the determinant occupies its specific site, it is not replicated, so that it will finally be diluted in the whole population. The integration site is supposed to be extra-chromosomal, possibly at a point in the cell membrane.

However, uncontrolled multiplication of R-factors can occur, especially in freshly infected cells (Lebek, 1963a, 1963b). This property, in addition to a very high transfer rate, leads to an epidemic spread of these factors throughout the sensitive population so that a high percentage of the bacteria is infected within a few hours. In this case, one must assume that replication of the R-factor proceeds much faster than multiplication of the host cell. However, repression reduces transferability to new cells. It is probable that in this state each bacterium carries only one copy of the complete R-factor, and thereafter replication is synchronous with the host chromosome.

The transferability of R-determinants can be lost by segregation of the R-factor during transfer, by segregation during multiplication in the host cell, or by elimination of the transfer factor by chemical agents such as acridine dyes or ethidium bromide, without damage to the R-determinant.

2. Stability of R-Factors

In vivo. It is difficult to assess the rate at which these factors can be lost *in vivo*, since cross-infections must occur constantly in man as well as in animals. This is due to the abuse of antibiotics which create the selective pressure necessary to maintain the predominance of the R-factor strains.

In vitro. WATANABE and FUKASAWA (1961 a) and MITSUHASHI et al. (1961 a; 1961 b) showed that R-factors can be eliminated by treatment with some acridine derivatives, but that the frequency of elimination is rather low. BOUANCHAUD et al. (1968), using ethidium bromide, eliminated between 12 and 100% of some resistance characters with several enterobacteria.

Ultraviolet irradiation can affect the *in vitro* stability of R-factors, but the observed elimination occurred mainly in resistance determinants (ANDERSON et al., 1968).

3. Molecular Nature of R-Factors

Like most genetic material, R-factors consist of DNA. This is established by the following data.

Acridine dyes inhibit replication of R-factors (MITSUHASHI et al., 1961 a; 1961 b) and it has been shown that these compounds induce insertions or deletions in DNA strands, being inserted between adjacent base-pairs in the DNA.

The incorporation of highly radioactive phosphorus and its subsequent decay normally destroys nucleic acids and inactivates R-factors (WATANABE and TAKAWA, quoted by WATANABE, 1963).

Ultracentrifugation in $CsCl_2$ density gradients with DNA extracted from cells carrying R-factors shows a satellite band in which the base composition differs from that of the chromosome. FALKOW et al. (1966) studied the G+C content of DNA of various R-factors carrying resistance to chloramphenicol, kanamycin, streptomycin, sulphonamides and tetracycline; this factor gave segregants with different resistance patterns during growth of the host cell. The determinant for chloramphenicol resistance has a high density, 1.716 $g \cdot cm^{-3}$, which corresponds to a G+C content of 56%; the density of the determinant for streptomycin and sulphonamide resistance lies between 1.711 and 1.714 $g \cdot cm^{-3}$, corresponding to a G+C content of 51–53%. The determinant for tetracycline resistance and the DNA of the transfer factor have a density of about 1.709 $g \cdot cm^{-3}$, a value which corresponds to 49% G+C; this value is equivalent to that of the *E. coli* chromosome. It is therefore unlikely that the genes responsible for the various resistances originated in the bacterium from which they were isolated; probably the determinants were incoporated into the R-factor genome during replication in the cytoplasm of different bacterial strains.

B. Mechanisms of Transfer

Transfer mechanisms of genetic material by cell-to-cell contact have stimulated many investigations. ANDERSON et al. (1957), on the basis of electron micrographs, suggested the formation of a conjugation bridge between F^+ (male) cells and F^- (female) cells. However, it has not been proved without doubt that this phenomenon is not an artefact. More recent work has shown that cells bearing an unrepressed F-factor always have some appendages on the surface. LOEB (1960) has described specific phages for cells bearing an F-factor. BRINTON et al. (1964) and BRINTON (1965) demonstrated that the receptor sites for the "male-specific" phages were located on special fimbriae, which are essential for the transfer of chromosomal genes. When the fimbriae of an Hfr strain are detached by violent stirring, the frequency of recombinants is drastically reduced. MEYNELL and DATTA (1966a, 1966b) have shown that F^- cells carrying some transfer factors were able to propagate these male-specific phages. They pointed out later that the fertility of these strains was related to the number of fimbriated cells (MEYNELL and DATTA, 1967). It is likely that these fimbriae play a major role in bacterial conjugation and are actively implicated in the transfer of resistance determinants. The diameter of the inner axial cylinder is about 20 to 25 A°, which could allow the passage of one DNA strand. Other workers believe that the fimbriae play a role only in the pairing of bacterial cells.

C. Classification of R-Factors

WATANABE (1963) divided the R-factors into two groups according to their effect on the fertility of strains of *E. coli* K12. Certain factors inhibit fertility, that is to say they reduce the frequency of transfer of chromosomal markers to F^- strains: they have been called fi^+ (for fertility inhibition); the factors which do not possess this property have been designated fi^-. ANDERSON (1966) demonstrated that these fi^- factors show, in contrast to the majority of fi^+ factors, a specific restrictive pattern of activity on typing phages of various *Salmonellae* especially those of *S. typhi*, *S. typhimurium* and *S. paratyphi* B. These restrictive properties allow the identification and characterization of many fi^- R-factors. A method has been set up allowing such a type of classification for fi^+ R-factors (PITTON and ANDERSON, 1970).

The degree of inhibition of lysis by male-specific phages produced by transfer factors can be determined by two simple methods:

a) Spot test. Application of a concentrated phage suspension to the surface of a sensitive indicator strain gives rapid results; however, the sensitivity of the method is limited.

b) Determination of the efficiency of plating (EOP). This gives more comparable results, allowing quantitative evaluation of the inhibitory effect of the various factors.

The transfer factors studied all originated from wild-type strains of *Salmonella* as shown in Table 1, and the male-specific phage used was μ2 (DETTORI et al., 1961). The results of these experiments are reported in Table 2.

Table 1. Origin of transfer factors and R-factors (donor strains)

ERL strain No	Phage type	Resistance of wild parental strain	Designation of R-factors or transfer factors	*fi* character
Salmonella serotype: *S. typhimurium*				
RT1	29	ASTSuFu	Δ	*fi*⁻
1R683	29	SSuTFu	683T	*fi*⁻
4R256	29	CKSSuTFu	256T	*fi*⁻
3R129	6	T	129	*fi*⁺
3R130	6	T	130	*fi*⁺
RT780	6	T	780	*fi*⁺
3R118	29	T	118	*fi*⁺
3R125	6	T	125	*fi*⁺
1R726	29	Sensitive	x_5[a]	*fi*⁺
4R256	29	CKSSuTFu	256C	*fi*⁺
Salmonella serotype: *S. paratyphi* B				
7R334	3a var. 4	ACKSSu	334	*fi*⁺

Resistances: A = ampicillin, S = streptomycin, C = chloramphenicol, T = tetracycline, K = kanamycin-neomycin, Su = sulphonamides, Fu = furazolidone.

[a] The transfer factor x_5 came from a wild-sensitive strain. It was associated with a determinant for ampicillin resistance by the triparental cross of Anderson (1965a, 1965b).

Table 2. Effect of various *fi*⁺ and *fi*⁻ factors in K12F⁺

Factor	*fi* character	Spot test	EOP μ2 (agar layer)	Plaque morphology (agar layer)
F		CL	1	Small clear
Δ[a]	*fi*⁻	CL	1	As F⁺
683T	*fi*⁻	CL	1	As F⁺
256T	*fi*⁻	CL	1	As F⁺
129	fi^{+1}	OL	1	Turbid
130	fi^{+1}	OL	4×10^{-1}	Turbid
780	fi^{+1}	OL	1	Turbid
118	fi^{+2}	VOL	4.5×10^{-2}	Very turbid
125	fi^{+2}	VOL	2.3×10^{-2}	Very turbid
x_5	fi^{+3}	VOL to —	$1.3\text{–}1.6 \times 10^{-1}$	Very turbid, difficult to count
256C	fi^{+4}	—	0	No detectable plaques
334	fi^{+4}	—	0	No detectable plaques

0 = No detectable plaques on agar layer, — = No detectable lysis on spot test.
[a] ANDERSON and LEWIS (1965b).

In this table, the first class of *fi*$^+$ factors has been designated *fi*$^{+1}$. When these factors are introduced into K12F$^+$ they do not completely inhibit lysis by phage µ2. The lysis is easily visible and turbid in spot and agar-layer tests. The second class is called *fi*$^{+2}$ because K12F$^+$ carrying such factors show very turbid lysis in spot tests, and the plaques are very turbid but countable in the EOP determinations. The third class is *fi*$^{+3}$ and these factors completely inhibit visible lysis in spot tests but lysis is still detectable in agar-layer tests, although the plaques are very faint. The factors producing maximum inhibition of lysis are designated *fi*$^{+4}$ and with these factors in K12F$^+$ both spot and agar-layer tests are negative with phage µ2.

Table 3. Effect of *fi*$^+$ transfer factors on K12HfrH

Factors	*fi*$^+$ character	Lysis by µ2		Transfer of *pro*
		Spot tests	agar layer	(K12HfrH = 1)
129	+1	—	Very turbid	7.7×10^{-3}
130	+1	—	Very turbid	3.9×10^{-2}
125	+2	—	Very turbid	1.7×10^{-2}
x_5	+3	—	—	2.7×10^{-4}
334	+4	—	—	8.8×10^{-4}
256C	+4	—	—	5.5×10^{-4}

To confirm fertility inhibition, the effect of the various *fi*$^+$ factors was examined by introducing them into K12HfrH and crossing the progeny with K12 F$^-$, according to the method of Clowes and Hayes (1968). The results, shown in Table 3, are calculated as a fraction of the transfer frequency of *pro* by HfrH.

These results provide sufficient information for a classification of the *fi*$^+$ transfer factors. These observations are interesting from both the genetic and epidemiological points of view, since it is possible to collect bacterial strains carrying identical factors and to determine the distribution of these factors in the different species of enterobacteria.

D. Transfer Frequencies

If the resistance determinants are accepted by a strain, then the frequency of transfer is the result of the interaction of several parameters; the principal ones, according to Anderson (1968), are:

a) The *recipient strain*. The various species of enterobacteria vary widely in their sensitivity to infection by an R-factor. Different *Salmonellae*, for example, can accept a given factor with very different efficiency if this factor is first isolated in a wild-type strain of *E. coli*. It is noteworthy that *S. typhi* (phage type A) is normally the best recipient for these *E. coli* factors.

b) The *donor strain*. Some strains are more efficient than others; this character depends essentially on the state of repression of the transfer factor in the given strain.

c) *The physiological state of the donor and recipient strains*. The transfer of resistance factors normally occurs at low frequency (about 10^{-5} per donor cell after a mating of one hour). However, if the cells are in optimum physiological conditions, the R-factors can spread very quickly in the recipient cells. This seems to be due to a derepression of the transfer factor.

d) *The rate of transfer of the TF alone*. This is normally constant under given conditions.

e) *The linkage between the TF and the R-determinant*. A reversible association results in a decrease of the transfer frequency of the complete R-factor; with a close association between them, the frequency of transfer of the complex will be high.

E. Origin of the R-Determinants

Although the resistance determinants need not have originated in the strain in which they are detected, they must have arisen, probably by mutation, on one of the following elements of the parental strain: chromosome, extrachromosomal part of the genome, or a transfer factor. No evidence has yet been submitted to support the hypothesis that R-determinants originated on transfer factors.

We shall try to demonstrate that the arguments in favour of a chromosomal origin of resistance determinants are certainly open to question. Firstly, the mechanisms of resistance mediated by R-factors are almost always different from those controlled by chromosomal genes; secondly, the high resistance levels that certain factors confer on their host cells would require a number of mutational steps and some of these high levels have never been obtained with mutants isolated *in vitro*; thirdly, it must be noted that, for certain antibiotics at least, the sensitive chromosomal gene is dominant over its resistant allele. Therefore, if this gene for sensitivity, to streptomycin for example, is in its normal form in the majority of cells carrying R-determinants for this drug, it has no effect on the function of these determinants. Finally, it must be assumed that, if determinants originate on the chromosome, a very specific homology would have to be supposed between transfer factors and resistance genes to allow a pick-up of these genes by the transfer factors. In the case of multiresistance, pick-up would have to be sequential and this seems rather unlikely.

Resistance Levels

Resistance levels conferred by R-factors upon their host depend on both the host cell and the R-determinant. WATANABE and FUKASAWA (1962) showed that a certain factor, which confers upon *Shigella* a streptomycin resistance of about 1000 μg/ml, gives a level of 10 μg/ml in a strain of *E. coli*. LEWIS (1967) showed that the streptomycin resistance of the recipient cells was much lower than that of the donor in a system of resistance transfer from *Shigella* to *E. coli* K12, but for chloramphenicol resistance he found the opposite reaction. In contrast, ANDERSON and LEWIS (1965a) observed that, in their *S. typhimurium* strain, the MIC for ampicillin was the same in the wild-type strain as in *E. coli* K12 into which the A determinant has been introduced. Moreover, some determinants confer resistance levels which seem to vary within the same strain.

SOMPOLINSKY et al. (1967), studying a transferable multiresistance for chloramphenicol, streptomycin and tetracycline, demonstrated that each of these resistances can be raised to a high level after exposure of the cells to an adequate concentration of each drug. When these determinants for high level resistance are eliminated by acridine, the cells become totally sensitive. The mutability towards high resistance seems to be a specific character of the determinant, independent of the chromosome, since these determinants for high levels are transferable to new sensitive cells.

F. Origin of R-Factors

ANDERSON (1965a, 1965b, 1967) suggested that some R-factors are formed by the union of an R-determinant and a transfer factor. To uphold this hypothesis, it was necessary to isolate wild-type strains carrying transfer factors lacking R-determinants, and conversely to isolate strains with non-transferable resistance determinants.

The useful technique described by ANDERSON (1965a), called the "mobilization test" or triple cross, is as follows:

$$A(x) \times B(r) \times C \rightarrow C(x\text{–}r).$$

The donor strain A is supposed to carry the transfer factor *x*; the intermediate strain B is a laboratory strain, carrying a non-transferable resistance determinant, normally obtained from a segregant during an interrupted mating mixture; the final recipient strain is a sensitive one, devoid of any resistance marker or any transfer factor. Thus this strain will acquire the complete factor (x–r) after recombination in strain B, only if the strain A carries the transfer factor x.

With this method, Anderson has been able to demonstrate that many wild-type sensitive strains of *S. typhimurium* carry transfer factors. He has also applied this test to show that some resistance determinants can recombine easily with a known transfer factor; the test can be done, as follows, with only two strains:

$$A(x) \times B(r) \rightarrow A(x) + B(x\text{–}r) \rightarrow A(x\text{–}r) + B(x\text{–}r).$$

This example explains the probable formation of complete R-factors in the wild-type strains of *Enterobacteriaceae.*

Formation of R-Factors for Tetracycline Resistance

The working hypothesis was based on the assumption that, if R-factors contain an element which has been picked up on the chromosome, it should be possible to recombine these genes with functional transfer factors. This would permit one to consider a chromosomal influence on the origin of transferable resistances. It has been claimed by LEBEK (1963c), MACUCH et al. (1967) and FRANKLIN (1967) that transferable tetracycline resistance can be evoked *in vitro* by exposing *Escherichia coli* to the drug.

CHUILT and PITTON (1969) isolated mutants resistant to between 20 and 100 μg/ml of tetracycline, from *E. coli* K12 and wild-type strains of *E. coli.* No transfer of tetracycline resistance from these strains could be demonstrated, even after treatment with mutagenic agents such as acridine derivatives or after UV induction.

Various wild-type *E. coli* or *Salmonellae* were used as transfer factor donor strains. All the strains carried transfer factors, some of which were fi^+ and some fi^-. Most of them carried resistance determinants for antibiotics other than tetracycline. One strain carried a determinant for colicinogeny.

The technique used to test transferability of the tetracycline resistance of the mutants was the resistance determinant mobilization test described by ANDERSON (1965a), using donor strains known to carry a transfer factor, an intermediate recipient – in this case the tetracycline-resistant *E. coli* mutants – and a final recipient, *S. typhi, S. typhimurium* and *S. paratyphi* B. The characteristics of the donor strains are summarized in Table 4.

It was not possible to mobilize the tetracycline resistance of the resistant mutants with any of the transfer factors used, that is, a tetracycline R-factor was not formed. Since the mating mixtures did not give a positive result, it can be supposed that the determinant producing tetracycline resistance can not be mobilized. It is probably chromosomal in location and thus inaccessible to the transfer factors.

These results support the hypothesis that transferable resistances did not originate in chromosomal mutants, but that their origin lies in another locus of the bacterial genome.

Other characteristics of the selected mutants are: mucoid state, reduced growth rate, decreased sensitivity to chloramphenicol. These characters are not normally detected in tetracycline-resistant wild-type strains of *E. coli.* It could be inferred from this that the resistance to tetracyclines in the wild-type strains is often extrachromosomal and that this type of cell is widely distributed among the wild pathogenic strains.

Table 4. Characteristics of the *E. coli* and *Salmonella* strains used for the crosses

Original strains	Derived strains	R-type	Other characteristics
E. coli R52		C	*fi*+, mucoid
	S. typhi (R52)	C	
	S. paratyphi B (R52)	C	
	S. typhimurium (R 52)	C	
	E. coli K12 F + (R52)	C	
	E. coli K12 F + (R52)	sensitive[a]	
	E. coli K12 F − (R52)	C	
E. coli R57		SC	*fi*+
	S. typhi (R57)	SC	
	S. paratyphi B (R57)	SC	
	S. typhimurium (R57)	SC	
E. coli R70		SC	*fi*+
	S. typhi (R70)	SC	
	S. paratyphi B (R70)	SC	
	S. typhimurium (R70)	SC	
E. coli R108		SC	*fi*+
	S. typhi (R108)	SC	
	S. paratyphi B (R108)	SC	
	S. typhimurium (R108)	SC	
E. coli R71		ASCT[b]	*fi*+
	S. typhi (R71)	SC	
	S. paratyphi B (R71)	SC	
E. coli R109		SC	*fi*
	S. typhi (R109)	SC	
E. coli R11		C	*fi*+ colicinogenic
E. coli K12 $F^- \Delta^+$ [c]		sensitive	*fi*−
E. coli K12 $F^- \Delta m^+$ [c]		sensitive	*fi*−
E. coli K12 (2144A) [c]		sensitive	*fi*+

[a] Strains obtained after a second series of crosses, i.e.: *Salmonella* × *E. coli* K12.

[b] The recipient *Salmonella* did not receive the T determinant (resistance to tetracycline).

[c] These strains were kindly provided by Dr. E. S. Anderson, Enteric Reference Laboratory, Central Public Health Laboratory, London.

Mechanisms of Resistance

In the second part of this paper the different mechanisms of drug resistance at the cellular and molecular levels will be discussed for each group of antibiotics. In each case, the mode and site of action of these drugs on the bacterial cell will be reviewed. The mechanism of drug-resistance will be contrasted with the mode of action.

IV. The Penicillins

All the penicillins now used in the treatment of bacterial infections are derivatives of 6-aminopenicillanic acid or 6-APA:

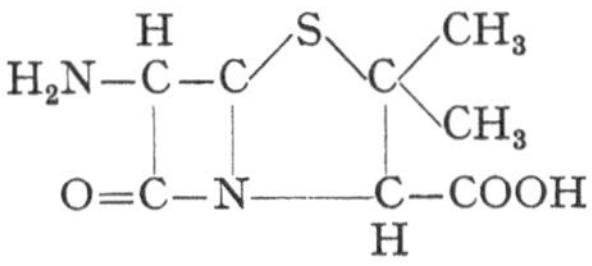

```
       H     S      CH3
  H2N—C—C ⁄   \ C ⁄
       |  |      | \CH3
    O=C—N———————C—COOH
                H
```

Whatever the substituent on the amino group on position 6, the mode of action is the same for each molecule carrying this chemical nucleus.

A. Mode of Action

Benzylpenicillin and all derivatives of 6-APA are bactericidal only to growing cells. They lead to the formation of spheroplasts, similar to those of lysozyme, but in contrast to the latter, which directly attacks the cell wall by its mucopeptidase activity, penicillin interferes with the formation of the mucopeptide. Deprived of their cell wall, bacteria lyse, and this accounts for the bactericidal action of the penicillins. The following experiments support this conclusion:

a) Cytological studies have shown that bacterial growth in the presence of sub-inhibitory concentrations of penicillin results in the formation of abnormal and swollen cells.

b) During growth of *S. aureus* in penicillin-containing media, it is possible to detect within the cell an accumulation of N-acetyl-muramic acid derivatives (PARK, 1952).

c) Growth of some bacteria in hypertonic conditions, for example 10% sucrose, and in presence of penicillin leads to formation of spheroplasts (LEDERBERG. 1956).

d) Penicillin inhibits incorporation into the cell wall of some specific amino acids, but does not inhibit the synthesis of cytoplasmic proteins (MANDELSTAM and ROGERS, 1959).

The current hypothesis is that penicillin prevents one of the final reactions in the formation of the cell wall. In staphylococci, these reactions could be as follows:

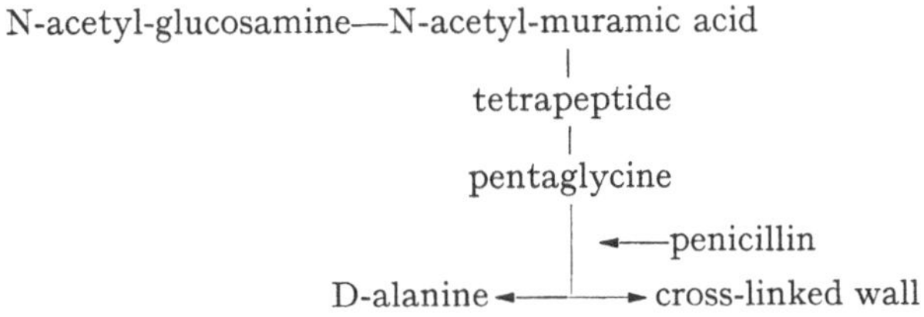

```
N-acetyl-glucosamine—N-acetyl-muramic acid
                          |
                     tetrapeptide
                          |
                     pentaglycine
                          |
                          |  ←—penicillin
             D-alanine ←——┴——→ cross-linked wall
```

In the presence of penicillin, polymerization is inhibited and no mucopeptide is formed.

B. Resistance to Penicillins

The synthesis of penicillinase by Gram-positive as well as by Gram-negative bacteria is an extensively studied mode of drug resistance. However, it is very unlikely that the acquisition of this type of resistance, involving modification or destruction of the drug, could result from one single mutation in a microorganism initially lacking this activity. The development of such a type of resistance would require the integration in the genome of genetic material containing at least one cistron able to code for synthesis of the drug-destroying enzyme.

1. Mode of Action of Penicillinase

The mode of action of penicillinase can be demonstrated by some simple techniques (Perret, 1954; Novick, 1962).

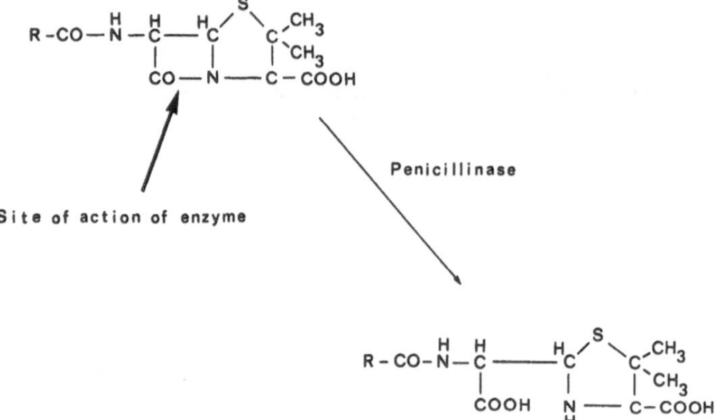

Fig. 1. Hydrolysis of penicillins by penicillinase. The enzymatic splitting of the β-lactam ring leads to the formation of penicilloic acid, biologically inactive

As shown in Fig. 1, the enzyme splits the β-lactam ring of penicillin, with the formation of penicilloic acid, which is biologically inactive. It is important to distinguish between this β-lactamase, which represents the most frequent type of penicillinase and the β-amidase (Batchelor et al., 1961). This enzyme hydrolyses the β-amido group on the carbon on position 6. This property is important for the preparation of 6-APA, used in the synthesis of the semi-synthetic penicillins. This enzyme is found in numerous enterobacteria and yeasts but it does not seem to be important in drug resistance from the clinical point of view, since the affinity of the enzyme for the substrate is rather low and because of its high pH optimum (Cole and Sutherland, 1966). The name penicillinase is taken to be synonymous with β-lactamase.

The problem of penicillinase synthesis must be considered in relation to the producing strain. It is important to differentiate between Gram-positive and Gram-negative microorganisms; Staphylococci and enterobacteria are typical examples of each. From the therapeutic point of view, they are the most important bacteria in which resistance to penicillins can be really dramatic.

2. Penicillin Resistance in Staphylococci

Drug-tolerant strains can be isolated *in vitro* in media containing penicillin. Resistance increases gradually and is lost when the cells are grown in a medium lacking the antibiotic. Such resistant strains normally show morphological modifications and a real extension of the lag phase. On solid media, it is often possible to detect small colourless colonies; the plasma-coagulase and α-toxin production are often lost. It seems that this type of resistance is of small importance in infectious diseases.

Strains of *S. aureus* able to destroy penicillins are almost always responsible for staphylococcal infections. They are similar to sensitive strains, except for their ability to produce the inactivating enzyme. This type of resistance is relatively stable; however, in *in vitro* subcultures it is possible to detect the emergence of an increasing proportion of sensitive variants which have completely lost their capacity to synthesize the enzyme.

Most attempts to demonstrate *in vitro* acquisition of the ability to produce penicillinase have failed in staphylococci (Barber, 1957, 1962).

With resistant staphylococci isolated in nature or in clinical specimens, penicillinase synthesis is always inducible (Geronimus and Cohen, 1957; Swallow and Sneath, 1962). The enzyme is formed in small amounts as long as the cells are not in contact with penicillin or a closely related compound which can act as a substrate. Moreover, it has been shown that these small amounts of basal penicillinase, formed by the inducible strains before the induction processes, are not sufficient to protect the cells against even low concentrations of penicillin. Induction is essential for the development of penicillin resistance.

Mutants with constitutive enzyme have been isolated, but they must be considered laboratory artefacts, since this type of mutant has never been isolated in nature (Pollock, 1962).

3. The Genetic Control of Penicillinase Production in Staphylococci

Novick (1963) observed that in certain strains of *S. aureus* the genetic determinants responsible for the control of penicillinase synthesis were incorporated in an extrachromosomal particle. This observation was confirmed by Harmon and Baldwin (1964) and Hashimoto et al. (1964).

The hypothesis was based on the following data:

a) The determinant responsible for the enzyme synthesis is irreversibly lost at a rather high frequency (about 10^{-3} per cell per generation). Some other genes giving resistance to mercuric chloride or to macrolides, which are in close relation with the penicillinase locus, are lost simultaneously.

b) The rate of formation of segregants which have lost all these properties is increased when the resistant population has been grown in a medium containing some non-mutagenic agents, such as high temperature, novobiocine, detergents, tetracyclines, etc.

The plasmids responsible for the control of penicillinase synthesis seem to behave as an independent genetic unit. The complete group is transduced as an unbroken unit, the loss of one character always producing loss of associated characters (Novick and Richmond, 1965).

Novick (1969) has proposed a classification of plasmids in *S. aureus*, reported in Table 5.

Table 5. Resistance plasmids in *S. aureus*

- Pen/Cad
  - PI
    - PenR EroS
    - PenR EroR
  - PII
    - PenR EroS
    - PenS EroS
- Other plasmids
  - Tet
  - Chl
  - (Kan)

This table shows the main resistance plasmids described in *S. aureus*. The Pen/Cad group belongs to a group of homologous plasmids which have been named "penicillinase plasmids", their principal function being the control of penicillinase synthesis. Since it is always present, the Cad character has been introduced because some plasmids of this group lack the penicillinase gene. PI and PII are two groups of incompatibility in which a more precise classification has been based on resistance to penicillin and erythromycin. Kan (resistance to kanamycin) has been placed within brackets to indicate the uncertainty of an extrachromosomal location.

Some data demonstrate that certain penicillinase determinants could be located on the chromosome. Poston (1966), studying the kinetic development of phenotypic expression and the effect of ultraviolet irradiation on frequency of transduction, thinks that the results favour a chromosomal location of the genes responsible for enzyme synthesis. Asheshov (1966a) describes three penicillin-resistant strains of *S. aureus*. This author has compared the frequencies of transduction before and after UV irradiation and concludes that, for one strain at least, the determinants are located on the chromosome, that for a second strain they are plasmid-linked, while the third could possess both chromosomal and extrachromosomal characters. In a study carried out on 50 wild-type strains of *S. aureus*, Asheshov (1966b) found 12 strains in which the determinants are extrachromosomally located.

It appears then that both locations can be found in the wild-type staphylococci.

Sawai et al. (1968b) have found two determinants responsible for penicillinase synthesis, one on the chromosome, the other on a plasmid, carrying identical properties in their structural and regulator genes. They conclude that a very close relationship exists between both types of determinant.

On the basis of all these observations, it may be considered probable that penicillinase-producing staphylococci, so frequent in hospitals, infect the patients by an epidemiological spread of resistant wild strains, rather than by the emergence of mutants arising from originally sensitive strains. The distribution of the resistant staphylococci is favoured by the following factors:

a) Staphylococci are part of the normal flora.

b) The elimination of sensitive bacteria in the treated patient creates ecological conditions in which the resistant strains can predominate.

c) Staphylocci are resistant enough to be transferred from one person to another.

d) This transmission is facilitated by a decrease in asepsis.

4. Penicillin Resistance in *Enterobacteriaceae*

The structure of the cell wall in Gram-negative bacilli is much more complex than in Gram-positive bacteria; this explains why no detailed study has been made of the action of penicillins in these organisms. Although these bacteria contain muramic acid, alanine and glutamic acid, which are specific components of the mucopeptide, the wall itself seems to contain only a small percentage of this compound. It was therefore necessary to determine the relationship between the mucopeptide and the physical properties of Gram-negative cell wall and to investigate whether the action of penicillin can be fully explained by the inhibition of mucopeptide synthesis.

The experiments described by Rogers (1962) have demonstrated that, in *E. coli* at least, the activity of penicillin is the same as in Gram-positive organisms.

As benzylpenicillin is only slightly inhibitory on the enterobacteria, studies with these organisms began with the first broad-spectrum semi-synthetic penicillin, the α-aminobenzylpenicillin or ampicillin (Rolinson and Stevens, 1961).

As far as resistance is concerned, it soon appeared that the situation was slightly different from that found in staphylococci. It was observed that when patients were treated with ampicillin for urinary tract infections by *E. coli*, the microorganisms isolated at the beginning of the treatment were sensitive to the drug, but that the bacteria isolated some days later were resistant to much higher concentrations of the antibiotic. Moreover, after exposure of the cells to the drug, the initially sensitive cells were able to produce *in vitro* variants resistant to high levels (Knox, 1962). It must be added that these mutants also showed increased resistance to other β-lactamines. The hydroxylamine test (KNOX and SMITH, 1962) proved that these strains could destroy ampicillin with formation of penicilloic acid.

Soon afterwards, SMITH (1963) obtained identical results. He started with a sensitive strain and, after subcultures in increasing concentrations of ampicillin, he isolated a resistant mutant able to synthesize about 50 times more enzyme than the initial strain. He also showed that the enzymatic activity was increased tenfold after disruption of the cells or after treatment with toluene; this implies the existence of a permeability barrier. Another important feature was the demonstration by SMITH that synthesis of penicillinase was not inducible, but constitutive. Treatment of the cells with 6-APA or phenoxybenzylpenicillin did not significantly increase the total content of enzyme.

The study of HAMILTON-MILLER (1963a) on penicillinase synthesis in *K. aerogenes* led to similar conclusions concerning location, inducibility and permeability phenomena.

PERCIVAL et al. (1963), after screening a number of enterobacteria (*E. coli, Proteus, Aerobacter* and *Pseudomonas*), pointed out that the resistance level of the majority of the naturally ampicillin-resistant strains was in direct relation to the quantity of penicillinase produced by the cells; in certain cases, however, drug-resistance was not associated with production of the enzyme.

It should be noted that the amount of penicillinase produced by resistant Gram-negative bacilli is small in comparison with that found in Gram-positive microorganisms. It is probable that in Gram-negative bacteria a very high resistance level is conferred by the synthesis of small amounts of enzyme and that the lipoprotein layer plays a major role in the prevention of drug penetration into the cell.

As far as inducibility is concerned, HAMILTON-MILLER (1963b) isolated two strains of *Proteus* in which the synthesis of enzyme was inducible. Induction can readily be demonstrated, in these strains, in which the initial levels of enzyme are too low to be detected.

The study of the substrate specificity of various penicillinases tends to show that the enzymes produced by Gram-negative bacteria present specificity patterns very different from enzymes synthesized by Gram-positive organisms; moreover, within the Gram-negative rods, each enzyme seems to possess a species-specific pattern (SMITH and HAMILTON-MILLER, 1963).

5. Location of Determinants for Penicillin Resistance in *Enterobacteriaceae*

It used to be thought that penicillinase synthesis was always controlled by chromosomal genes until ANDERSON and DATTA (1965) showed not only the emergence of wild-type strains of *S. typhimurium* resistant to ampicillin, but also the fact that this resistance was transferable to sensitive strains of *E. coli*.

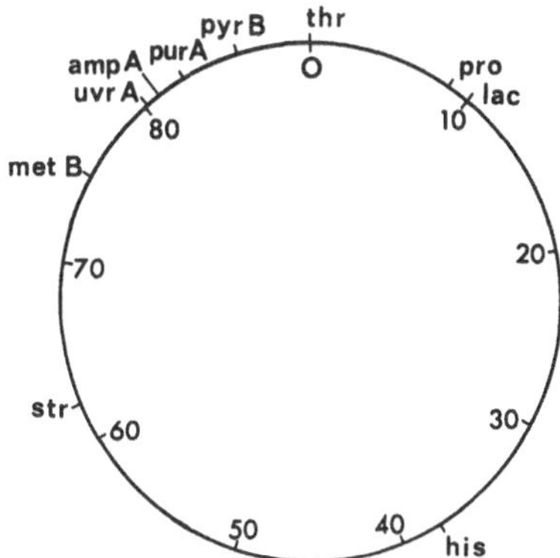

Fig. 2. Genetic map of *E. coli* K12. The genetic map shown here corresponds to an HfrHayes strain, with the origin located near the locus *thr*. The sequence of transfer of chromosomal genes is thus: *thr–pro–lac*

a) Chromosomal Location

Most of the work on this subject has been done by Eriksson-Grennberg et al. (1965). Starting from some strains of *E. coli*, it is possible to isolate one-step mutants resistant to low levels of ampicillin, the MIC being about 10 μg/ml. It must be pointed out, however, that with some other strains the authors have not been able to detect such mutants. The gene, named *ampA*, responsible for this first resistance level seems to be located on the chromosome in the region 80 to 85 (according to an Hfr Hayes strain) and the exact sequence of the genes could be *uvrA–ampA–purA–pyrB*; the chromosomal map would be as illustrated in Fig. 2. The resistance due to the mutation on this gene is mediated by the synthesis of a splitting enzyme.

A second type of mutation, named *ampB*, can double the resistance level of the cell carrying the *ampA* character, but without affecting enzyme production. It seems that this *ampB* mutation modifies some component of the cell membrane, but it is still impossible to decide whether the mutation affects a structural or a regulator gene. Cells carrying this mutation have decreased osmotic resistance, especially after EDTA treatment, they produce mucoid colonies and the efficiency of plating with phage T4 is strongly reduced; the cells are tolerant to colicins E2 and E3, although their adsorption is normal. This *ampB* mutation also induces increased resistance to other antibiotics active on the cell wall, as well as to streptomycin and chloramphenicol. The *ampB* phenotype seems to be due to a polygenic system, one locus being located near *gal* and two loci in the *trp* region (Eriksson-Grennberg, 1968; Nordström et al., 1968; Boman et al., 1968).

b) Specificity of Chromosome-Controlled Penicillinases

Some of our knowledge on this subject can be summarized (Sawai et al., 1968a), as shown in Fig. 3. These authors using 28 different strains of entero-

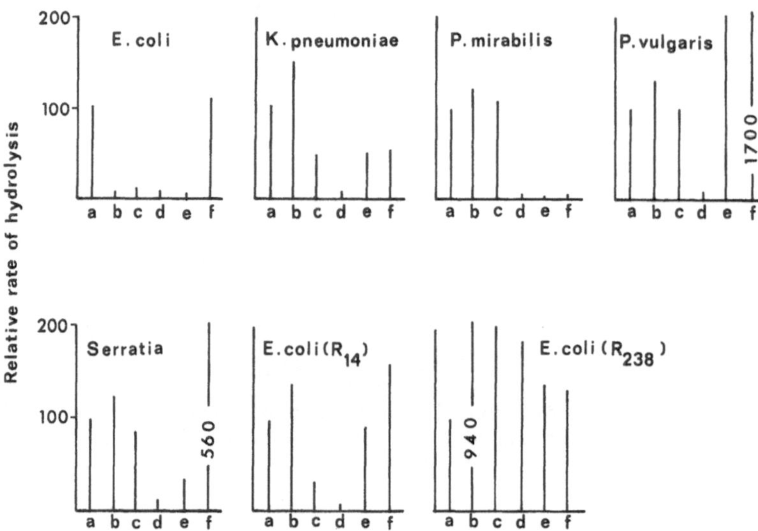

Fig. 3. Substrate profiles of β-lactamases of some Gram-negative bacilli. These β-lactamases are produced by various ampicillin-resistant enteric bacteria and by strains of *E. coli* carrying transferable R-factors. The activity on the different substrates of the various "species-specific" enzymes and that of β-lactamases controlled by the two R-factors is expressed graphically. The ordinate gives the relative rate of hydrolysis for each substrate in relation to that of benzylpenicillin, considered as 100; the different substrates on the abscissa are: *a* benzylpenicillin, *b* ampicillin, *c* pheneticillin, *d* cloxacillin, *e* 6-APA, *f* cephaloridin

bacteria (*Escherichia, Klebsiella, Aerobacter, Serratia* and *Proteus*) were unable to demonstrate any transferable character. β-lactamases are divided into two groups or types of enzyme, penicillinases and cephalosporinases, classified according to their specific activity on different substrates, with frequent overlapping. In contrast to the β-lactamases found in Gram-positive organisms, cephalosporinases seem to be more frequent than penicillinases in Gram-negative bacilli (Fleming et al., 1963). It must be pointed out that if the penicillinase-type enzyme is normally constitutive, the cephalosporinase-type is generally inducible and this is in accordance with the results of Hennessey (1967), Jago et al., (1963) and Sabath et al., (1965). The data of Sawai show that Gram-negative bacilli producing β-lactamases are highly resistant to penicillins and cephalosporins; this implies that in the enterobacteria production of enzyme is the most important mechanism of resistance to these antibiotics.

More recently, Jack and Richmond (1970) have studied 46 strains of enteric bacteria. They detected eight different types of β-lactamases, classified according to their substrate profile, sensitivity to p-chloromercuribenzoate, cloxacillin inhibition, reaction with one antiserum and charge properties. They found that the enzymes range from predominantly cephalosporinase activity to a more specifically penicillinase activity. However, the majority of the strains studied could synthesize an enzyme almost equally active on penicillins and cephalosporins.

c) Extrachromosomal Location

The work of ANDERSON and DATTA (1965) first demonstrated the emergence of ampicillin-resistant strains of *S. typhimurium*. For each strain they studied, the ampicillin resistance was part of a multiple resistance and was easily transferable to sensitive bacteria. The epidemiological data showed that the majority of these strains were the progeny of one clone and that they probably had only one ancestor. The fact that R-factors could mediate the production of a penicillinase was interesting, since it was possible for the first time to predict an analogy between Gram-negative rods and staphylococci, since it was suspected that in the latter a plasmid was responsible for penicillin resistance.

The specificity of the enzyme plays a major role in the understanding of the control mechanism. Since cells which lack a transferable determinant are capable of producing a certain amount of enzyme, one may suppose that the extrachromosomal elements derepress the chromosomal gene which is normally under repression. The enzymatic activity would then be increased by derepression. Nevertheless, the cell-free extracts of R^+ strains are very different from those obtained with penicillinase-producing mutants, not only in their high enzymatic activity, but also in the specificity patterns on different substrates. It can thus be concluded that the R^+ strains synthesize one (or more) penicillinase which cannot be detected in the R^- strains (DATTA and KONTOMICHALOU, 1965).

The β-lactamases whose synthesis is controlled by extrachromosomal determinants are constitutive enzymes and can be classified in the penicillinase-type according to their specific activity on substrates containing the thiazolidin-β-lactam ring.

d) Chemical Nature of Penicillinases

DATTA and RICHMOND (1966) were the first to isolate and purify a penicillinase produced by an *E.coli* strain carrying an R-factor. The protein which they obtained has a molecular weight of 16700, determined by equilibrium-ultracentrifugation, its pH optimum is between 5.7 and 7.0.

One of the main differences between this enzyme and the penicillinase from Gram-positive bacteria is its molecular weight, which is about half. On the other hand, the enzyme described by DATTA is very different from that of Gram-positive cells, since the amount of penicillinase produced by a derepressed culture expressed in enzyme units per mg of dry cells is about one tenth of that obtained with Gram-positive organisms.

Recently, LINDQUIST and NORDSTRÖM (1970) purified another penicillinase, mediated by an R-factor, for which the molecular weight is about 22000, determined by gel-filtration on Sephadex. This enzyme does not seem to be related to the enzyme mediated by the *ampA* gene, since the latter has a molecular weight of 29000, its substrate profile is very different and there is no immunological cross-reaction between the two enzymes (LINDSTRÖM et al., 1970).

6. Origin of Determinants Responsible for the Control for β-Lactamase Synthesis

If one accepts the pick-up of a gene on the chromosome by a transfer factor, then one must assume that determinants responsible for the enzyme synthesis have arisen in a limited number of species of *Enterobacteriaceae*. But the clear-cut distinctions between the specificity of these enzymes (as shown in Fig. 3) suggest that there is a difference, on the one hand between the R-determinants carried on the various R-factors, and on the other hand between these and the chromosomal genes.

Thus, in the Gram-negative bacilli the genes responsible for β-lactamase production can be located either on the chromosome or on an R-factor, whether transferable or not. The degree of infectivity of these factors in enterobacteria is so high that it seems impossible to consider the penicillinases as "species-specific" as it has not so far been demonstrated that their synthesis is controlled by chromosomal genes.

7. Phenotypic Expression of R-Determinants

When R-factors carrying determinants for β-lactamases are introduced in certain strains of Gram-negative rods (*Escherichia*, *Serratia* or *Aerobacter*), it can be assumed that the amount of enzyme produced is identical for each of these strains; but when the same determinants are transferred to certain strains of *Proteus* the amount produced is only a tenth or a hundredth of that obtained with the other strains (SMITH, 1969).

It may be supposed that gene dosage effect (JACOB et al., 1960) is involved in these cells. A major objection to this hypothesis is based on the observations of JACOB and MONOD (1961) who found that the production of β-galactosidase directed by an F-lac episome is two to three times higher than that obtained with a normal haploid bacteria, in which the *lac* gene is chromosomally located. The difference could be due to the fact that the replication of the plasmid is completed before that of the bacterial chromosome. As a proof of a gene dosage effect, it would be necessary to measure directly the amount of genetic material belonging to the extrachromosomal element.

8. Enzyme Location in the Cell

NEU and CHOU (1967) have shown that osmotic shock could release the surface enzymes (periplasmic enzymes) from the majority of *Enterobacteri-*

aceae. The technique allows an almost complete release of β-lactamase in *E. coli* and *S. typhimurium* strains in which the enzyme synthesis is controlled by a cytoplasmic R-determinant.

NEU (1969) described two main groups of β-lactamases, distinguished by their cellular location. The first group is formed by the periplasmic enzymes; these are easily released by osmotic shock, the resistance level of the cells producing this enzyme is very high, the MIC being about 1000 μg/ml of ampicillin, and the enzyme synthesis would be controlled by a plasmid. The second group includes the enzymes which are firmly bound to the bacterial cytoplasm and which are not released by osmotic shock but only by ultrasonic treatment of the cells; the resistance level, between 20 and 200 μg/ml of ampicillin, is much lower than in the first group; the control of enzyme synthesis would depend upon chromosomal genes.

It is interesting to note that synergistic action between penicillins such as ampicillin and oxacillin could be explained by a competitive inhibition of the enzyme due to the penicillinase-resistant derivative. Unfortunately, synergy is found only with the low-level resistant strains. It can be postulated that, in high-level resistant cells synthesizing a periplasmic β-lactamase, the enzyme has a great advantage since it can inactivate the antibiotic before it enters the cell and is recognized by the transpeptidase-enzymes mediating cell-wall synthesis.

9. Penicillinase Synthesis in Some Gram-Negative Bacilli

The problem of resistance to penicillins being important in the treatment of infections due to Gram-negative bacilli, the mode of genetic control of enzyme synthesis was determined in a certain number of *E. coli* and *Salmonellae* strains resistant to ampicillin (PITTON, unpublished results).

Resistance spectra and transferability of ampicillin resistance are shown in Table 6.

Table 6. Characteristics of the strains of *E. coli* and *Salmonella*

Number of strains		Resistance pattern	Transferability		
			positive		Not demonstrated
E. coli	*Salmonella*		*E. coli*	*Salmonella*	
6		A	0		6
3	2	AS	1	2	2
5	2	AT	4	2	1
2		AC	0		2
5	1	ASC	2	1	3
3	1	AST	2	1	1
6		ACT	3		3
13	1	ASCT	3	1	10
43	7		15	7	28

The ability to inactivate ampicillin in a cellular system has been studied by adding the drug to growing cultures and incubating for 6 hours; the residual antibiotic activity was determined by a microbiological assay. The results obtained are summarized in Table 7.

Table 7. Results of inactivation tests with intact cells

Rate of inactivation	Number of strains, Transferability		Total
	positive	not demonstrated	
Strong inactivation	3	1	4
Total inactivation	19	13	32
Partial inactivation	0	14	14

A total inactivation corresponds to a residual antibacterial activity of less than 10%; strong inactivation to a residual activity of between 10 and 50% and partial inactivation to a residual activity of between 50 and 80%.

It can be seen that all the strains for which the ampicillin character is transferable give a residual antibacterial activity of less than 50%. These strains seem to produce a high level of penicillinase.

Within the strains which did not transfer their resistance, two groups can be distinguished:

a) The first group contains cells for which the inactivation character behaves like it does in strains carrying a transferable determinant. It can be postulated that these strains have an extrachromosomal R-determinant for ampicillin-resistance but that the transfer factor is lacking.

b) The second group, showing weak enzymatic activity, contains strains in which no transfer mechanism has been demonstrated. It may be considered that in these bacteria the control of penicillinase synthesis is chromosomal.

In order to specify the character of some of these strains, inactivation tests with cell-free extracts have been carried out. To prepare the cell-free extracts, two methods were used:

a) Release of the enzyme by osmotic shock according to NEU and CHOU (1967).

b) Total release of the protein content of the bacteria by alumina grinding of frozen cells.

The supernatant of the cultures was also used to test enzymatic activity.

The typical inactivation curves obtained are shown in Fig. 4 and 5.

The results for all the strains studied show that the 7 strains of *Salmonella* whose resistance character is transferable have identical behaviour; it is assumed that the enzyme synthesis is controlled by a transferable plasmid and that the enzyme is periplasmic and easily released, even in culture medium.

In the 13 strains of *E. coli* the situation is a little more complicated; for five strains the results are identical to those obtained with the *Salmonellae*.

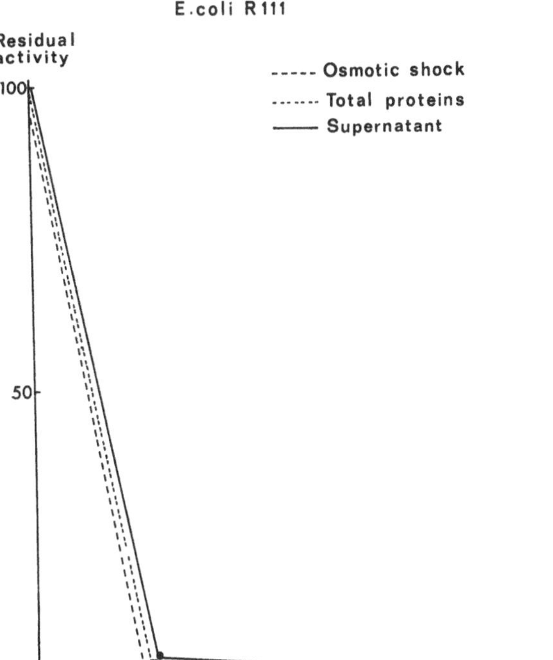

Fig. 4. Typical curve of total inactivation. The strain of *E. coli* R111 is a wild-type strain, carrying a multi-resistant R-factor (with resistance to ampicillin, streptomycin and chloramphenicol). All these characters are transferable to *S. typhi* and *S. typhimurium*, but are not transferable to *S. paratyphi* B. In the test described, the final concentration of total proteins is 1 µg/ml. This curve corresponds to a periplasmic penicillinase controlled by an extrachromosomal R-determinant

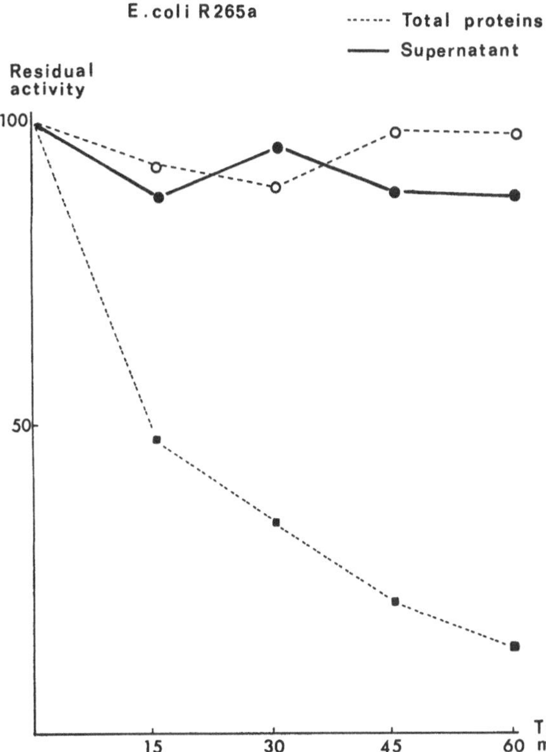

Fig. 5. Typical curve of partial inactivation. The strain of *E. coli* R265a is a wild-type strain resistant to ampicillin, streptomycin and chloramphenicol. Although the S and C determinants are transferable to *Salmonellae*, the transfer of the A character has never been detected. In the test, the final concentration of total proteins is 150 µg/ml. This curve corresponds to an intra-cellular penicillinase, not liberated by osmotic shock and probably synthesized under the control of chromosomal genes

In one of these strains the transferability of resistance has not been demonstrated, but the four other strains are positive in transfer tests; it can be postulated that for these 5 strains the enzyme synthesis is plasmid-controlled. A second group is formed by 6 strains for which the supernatant of cultures contains no detectable enzymatic activity in the tests. Accordingly, the cell-free extracts obtained after osmotic shock are also inactive; if the tests are done with total proteins, a complete but slow inactivation of the drug is obtained. In these strains penicillinase synthesis is not mediated by a plasmid like that found with the previous strains. The control could be exerted by chromosomal genes and they could be real mutants. Two strains are an exception, but they are mucoid and it can be supposed that the high mucopolysaccharide formation inhibits the release of enzyme either spontaneously in the culture medium or in preparations obtained after osmotic shock.

10. Characterization of Some Plasmid-Linked Penicillinases of Enterobacteria

In order to obtain more information on the mode of resistance to β-lactamines in Gram-negative bacilli, three penicillinases were isolated and purified. Their synthesis is mediated by transferable R-factors. The strains used were *E. coli* R111, *E. coli* R453 and *S. typhimurium* RS133; they are all wild-type strains isolated from clinical specimens.

The isolation and purification of the enzymes was done by the following methods.

Wet cells were ground with aluminum oxide and then dispersed in a phosphate buffer; a first partial purification was obtained by ammonium sulphate fractionation. After dialysis, the material was loaded onto a DEAE-cellulose column which was washed with phosphate buffer. The enzyme was eluted with an exponential gradient of sodium chloride; the enzymatic activity of the eluate was determined according to NOVICK (1962). The active fractions were dialysed and loaded on a G-75 Sephadex column and eluted with phosphate buffer; the active fractions were again dialysed, freeze-dried and dissolved in one tenth of their initial volume.

Four tests were used to characterize and differentiate the three enzymes.

a) Sodium Dodecylsulfate-Disc-Gel Electrophoresis

The electrophoretic analysis was done in SDS-gels, after treatment of the proteins by the detergent and mercaptoethanol. As shown in Fig. 6, the three enzymes have the same mobility under these conditions and can be considered relatively pure.

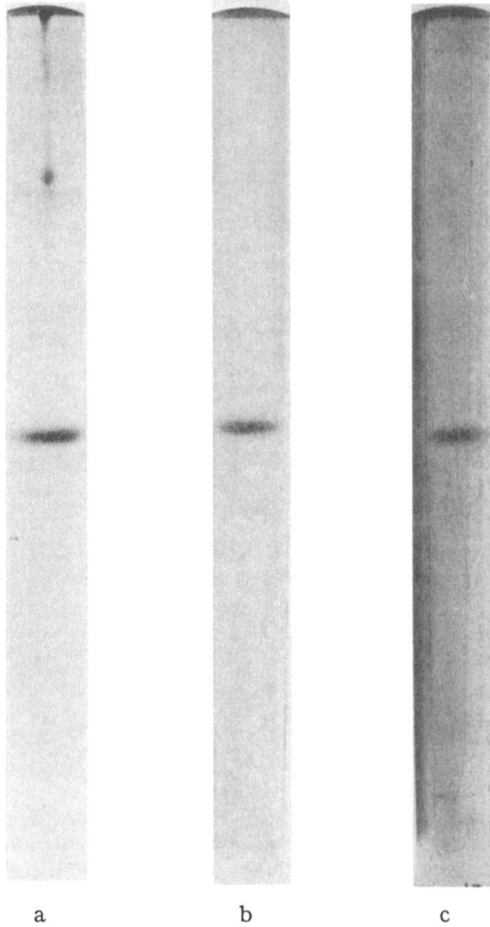

a b c

Fig. 6a–c. Sodium dodecylsulfate disc-gel electrophoresis. a Penicillinase from *E. coli* R111. b Penicillinase from *E. coli* R453. c Penicillinase from *S. typhimurium* RS133. The anode is at the bottom of the gels

b) Molecular Weight

The molecular weight was determined by SDS-gel electrophoresis according to the method of WEBER and OSBORN (1969) and the calculations were done according to AHMAD-ZADEH and PIGUET (to be published), using L-glutamate-dehydrogenase, ovalbumin, alcohol-dehydrogenase and myoglobin as standard proteins. The molecular weight was found to be the same for the three enzymes, about 26600.

c) Immunochemical Properties

Rabbit antisera were prepared with the R111 and the R453 enzymes and adsorbed with proteins obtained from ampicillin sensitive *E. coli* K12 F^-.

The immunochemical characteristics of the enzymes, as determined by immuno-electrophoresis (IEP), are shown in Fig. 7.

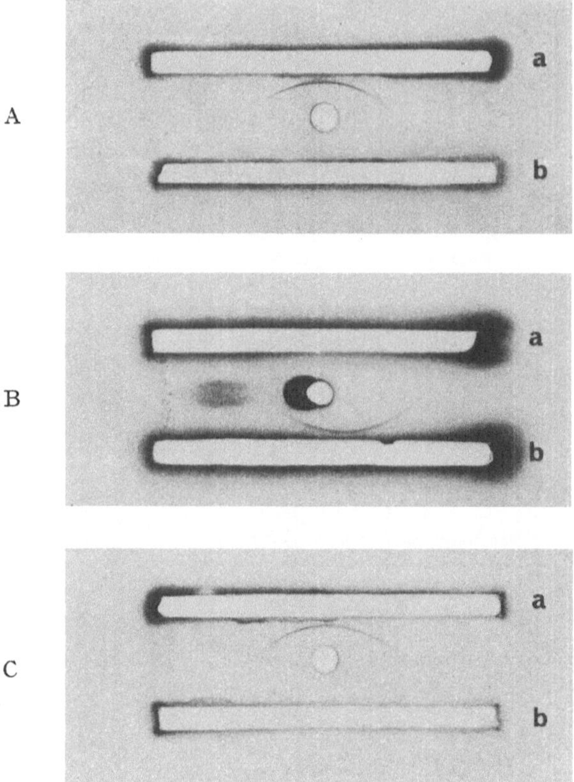

Fig. 7A–C. Immuno-electrophoretic analysis. A Penicillinase from *E. coli* R111. B Penicillinase from *E. coli* R453. C Penicillinase from *S. typhimurium* RS133. *a* Antiserum (R111). *b* Antiserum (R453). The anode is on the left

These results corroborate the elution properties of the enzymes, since the R111 and RS133 enzymes, presenting a week anodic migration in IEP, are eluted on DEAE-cellulose with a low concentration of sodium chloride, whereas the R453 enzyme, which shows a cathodic migration in IEP, is eluted in the washing volume on DEAE-cellulose. Moreover, it can be seen on the immunoelectrophoresis figures that the enzymes from R111 and RS133 are identical as far as their immunological reactions are concerned.

d) Identity of the Enzymes

To check the identity of the three enzymes, the following tests were done:

a) The antisera adsorbed with proteins obtained from sensitive variants of the parental strain which were selected by replica-plating show the same precipitation lines when they react with the proteins extracted from the resistant cells.

b) The enzyme preparations were placed on polyacrylamide gels prepared in phosphate buffer. After electrophoresis, the gels were cut longitudinally;

one half was placed on a Petri dish containing nutrient agar with *B. subtilis* and 10 µg/ml of ampicillin; the other half was placed on agar and allowed to react with the antiserum. The results show that the enzymatic activity is located at exactly the same place as the precipitation line obtained with the specific anti-serum. Moreover, the immunological tests done with both anti-sera have never shown a cross-reaction between the two enzymes. Thus, although the specific activity of the enzymes is very close when tested by conventional biochemical methods, it is certain that they differ in some of their immunological determinants and can easily be differentiated by simple immunochemical techniques.

After that, the proteins of 133 strains of Gram-negative bacilli were used in IEP and Ouchterlony's tests. The results are summarized in Table 8.

Table 8. Results of immuno-electrophoretic and Ouchterlony's test with Gram-negative bacilli

Number of strains studied		Reaction with anti-serum R111		Reaction with anti-serum R453	
		positive	negative	positive	negative
Escherichia coli	63	45	18	5	58
Salmonella	17	13	4	1	16
A. cloaca	4	1	3	1	3
Klebsiellae	33	5	28	10	23
Proteus	10	2	8	2	8
Acinetobacter	4	0	4	0	4
Providencia	2	2	0	0	2
Total	133	68	65	19	114

It can be seen that, of the 133 strains, 87 give a positive reaction with one or the other antiserum; this means that at least 65 % of the Gram-negative bacilli studied owe their ampicillin-resistance to the presence of an R-determinant. But the figure is more striking if one considers the *E. coli* and *Salmonella* strains, since the proportion is respectively 71 % and 82 % of plasmid-controlled penicillinase in these strains. It is interesting to note that the R111-type of enzyme is more frequent in these bacteria than the R453-type; an opposite reaction is shown by the *Klebsiella* strains in which the R453-type is twice as frequent as the R111-type.

V. The Aminoglucosides

The main antibiotics of this group (streptomycin, neomycin and kanamycin) have a very similar chemical structure. They show a similar antibacterial spectrum and the mutants normally present a relatively specific resistance, although some cross-resistances can be detected, especially between neomycin and kanamycin. Two other recent aminoglucosides, paromomycin and gentamycin, will not be included in this study, since genetic and biochemical data for them are not available.

Streptomycin

A. Mode of Action

This drug is bactericidal, after a lag period of a few minutes; but, unlike the penicillins, it produces practically no lysis of bacterial cells. A short period of growth is essential to its action, and the activity of the drug stops when protein synthesis is hindered by chloramphenicol or deprivation in an essential amino acid in an auxotrophic strain, for example.

Streptomycin activity is very sensitive to anaerobic conditions; it is without any effect on obligate anaerobes such as *Clostridia*. The reason for this lack of activity remains obscure.

The concentrations required for a lethal effect are increased drastically by lowering the pH, or with high concentrations of certain ions such as Mg^{++} or phosphate. These data suggest that the ionic strength plays a very important role in the binding of the antibiotic on the cell or in its relation with the specific intracellular site of action.

On adding streptomycin to a bacterial culture, many effects can be observed

a) One may first notice an inhibition of protein synthesis synchronous with the beginning of bactericidal action.

b) Later, the antibiotic inhibits some mechanisms of respiration, and after a transient stimulation, the synthesis of nucleic acids (DUBIN, 1964).

c) Finally, the drug leads to a degradation of cell membrane, with a loss of intracellular metabolites such as nucleotides and amino acids (BROCK, 1966).

Site of Action

This was determined by studying the streptomycin-dependence. It is now known that sensitivity and resistance (chromosomally determined) to this antibiotic, as well as dependence on it, are due to allelic forms of one gene. SPOTTS and STANIER (1961) showed that protein synthesis was reduced in the early stages of the growth cycle in dependent cells in absence of the drug. This reaction is identical to that observed in sensitive cells treated with streptomycin. These authors suggested that the antibiotic acts directly on ribosomes. Ribosomes of dependent mutants are functional in protein synthesis only in presence of the antibiotic: those from sensitive cells are inhibited by it and ribosomes of resistant alleles are insensitive to the action of the drug. For streptomycin, it is impossible to separate the study of the control of sensitivity, resistance or dependence when these reactions are under the control of a chromosomal gene.

nek aroE spc str pabA malA

Fig. 8. Chromosomal mapping of genes responsible for reactions to aminoglucosides. All these genes are located between the region 64 min (*aroE*) and 66 min (*malA*). The *str* gene controls reactions to streptomycin and dihydrostreptomycin, the locus *spc* corresponds to spectinomycin and the gene *nek* is responsible for reactions to neomycin and kanamycin

Location of chromosomal genes. The locus responsible for the control of reactions to streptomycin is located in the 64 min region, that is to say between the genes *aroE* and *argD* as shown in Fig. 8 (TAYLOR and TROTTER, 1967).

The studies of the mechanism of action of streptomycin have shown the exact site of action.

The results can be summarized as follows:

a) By association in cell-free systems of ribosomes and enzymes obtained from resistant and sensitive strains, the study of protein synthesis shows that only ribosomes are concerned in the determination of sensitivity or resistance (FLAKS et al., 1962; SPEYER et al., 1962).

b) The reassociation of 50S and 30S ribosomal subunits, isolated from sensitive and resistant strains, or from sensitive and dependent strains, gives ribosomes which are sensitive, resistant or dependent, the reaction being determined solely by the origin of the 30S subunits (COX et al., 1964; LIKOVER and KURLAND, 1967).

c) Reconstitution of 30S subunits from core particles and split proteins, gives sensitive or resistant subunits, according to the origin of the core particles (STAEHELIN and MESELSON, 1966; TRAUB et al., 1966).

d) 30S subunits sensitive or resistant to streptomycin can be formed by taking 16S RNA and soluble proteins obtained from sensitive or resistant strains (TRAUB and NOMURA, 1968).

As it is now possible to separate nearly all the proteins of the 30S subunits, methods of reconstitution have allowed the identification of the protein responsible for the phenotypic expression of the chromosomal mutation. OZAKI et al. (1969) have given the name of P10 to the protein, which is controlled by the *str* locus and determines the sensitivity of the 30S subunits to the inhibitory effect of streptomycin.

Some phenomena of antagonism between antibiotics can thus be explained more precisely; the effect of chloramphenicol on streptomycin activity can be explained by the fact that the latter can bind irreversibly only on ribosomes which are not linked with m-RNA. The stability of m-RNA *in vitro* is probably a consequence of the stability of the linkage between m-RNA and the ribosomes (GROS et al., 1963). Chloramphenicol inhibits degradation of the m-RNA. It is probable that the antagonistic action of chloramphenicol on streptomycin is due to a stabilisation of the m-RNA, the consequence being that the specific site of action of streptomycin is not free for a binding.

B. Resistance to Aminoglucosides

1. Chromosomal Resistance to Streptomycin

The results obtained by studying the *in vitro* polypeptide synthesis support the hypothesis that a ribosome has only one site on which streptomycin can bind to disturb protein synthesis. Every mutation which affects this site or which modifies it so that the antibiotic cannot exert its action must be considered as a mutation towards resistance. The isolation of one-step resistant mutants (MIC of about 1 mg/ml) agrees with the hypothesis of only one site of action.

It is important to point out that the character of chromosomal resistance is a recessive one (Lederberg, 1951). This phenomenon can be explained by the formation of polyribosomes, since in a heterozygote such elements must contain a mixture of sensitive and resistant ribosomes; the effect of antibiotic on any sensitive particle will result in the synthesis of an abnormal protein.

2. Other Aminoglucosides

The mode of action and the mode of resistance to kanamycin and neomycin are only partially explained. These two drugs also induce miscoding of the m-RNA and inhibition of protein synthesis. Normally, the mistakes they cause are more important than those due to streptomycin; on the other hand, the other aminoglucosides do not show a simple stoichiometric relationship between the drug and the ribosomes. This could imply that they interact with more than one ribosomal site or with more than one component active in the protein synthesis complex (Davies and Davis, 1968).

3. Resistance Levels

The resistance levels to aminoglucosides mediated by chromosomal mutations depend largely upon the antibiotic considered. Although mutants resistant to very high concentrations of streptomycin (such as 10 mg/ml) can be isolated, the one-step resistance levels to kanamycin or neomycin reach only 200 μg/ml for the first drug and 30 μg/ml for the second.

4. Other Phenotypic Reactions

The mutants resistant to aminoglucosides present some modifications in their phenotype:

a) Restriction and modification of some phages (Lederberg, 1957; Couturier et al., 1964).

b) Reduction or even loss of drug-dependent phenotypic suppression normally shown with the sensitive parent strain (Anderson et al., 1965; Apirion and Schlessinger, 1967a).

c) Reduction or elimination of the properties of a pre-existing genetic suppressor (Lederberg et al., 1964; Davies, 1966; Kuwano et al., 1968).

5. Specificity of Resistance

It seems that a very high specificity appears for mutations affecting resistance to aminoglucosides. Only a few cross-resistances are shown, except for closely related derivatives. Thus, *E. coli* mutants selected for a high resistance level to streptomycin or dihydrostreptomycin show a complete cross-resistance to both drugs, but they are normally sensitive to other aminoglucosides. Moreover, resistance to neomycin is usually associated with a certain degree of resistance to kanamycin and conversely (APIRION and SCHLESSINGER, 1968). In the *nek*+ mutants, resistant to neomycin and kanamycin, the ribosomes are modified in comparison with those of the parent strain. When they are used in cell-free extracts where synthesis of the polypeptide is directed by poly-C or poly-A, the misreading in the presence of the antibiotics is decreased in comparison to that obtained with ribosomes isolated from the sensitive strains, and the inhibition of protein synthesis due to the drugs is reduced.

Conjugation and transduction show that these mutations are closely linked or even allelic and that the genes responsible are located in the same region as the gene for streptomycin resistance. As for the latter, the *nek*+ character is recessive on the wild sensitive allele and, as with streptomycin-resistant mutants, pleiotropic effects can be observed. Increasing resistance to low concentrations of streptomycin or spectinomycin but decreased resistance to chloramphenicol are detected; this effect could be explained by the fact that, although *nek*+ mutants modify the 30S subunits, a mutation affecting one of the ribosomal subunits may act indirectly on the function of the other subunit, in this case the 50S on which chloramphenicol is active.

6. Dependence

It is necessary to point out briefly the mechanism of dependence. With some bacteria it is possible to isolate mutants with an absolute requirement for such drugs as streptomycin, neomycin or paromomycin.

Dependent cells are found among the resistant mutants obtained by the plating of a sensitive strain on a medium containing the antibiotic. When these dependent bacteria are subcultured without the drug (SPOTTS, 1962), synthesis of macromolecules stops, with a decrease of the activity of protein synthesis, but without any effect on nucleic acid synthesis.

The genetic data demonstrate that all these characters are allelic and that dependent mutants form a class of resistant mutants (APIRION and SCHLESSINGER, 1967b; LUZZATTO et al., 1968; APIRION et al., 1969).

7. Resistance to Aminoglucosides Mediated by Extrachromosomal Elements

Strains of enterobacteria carrying transferable streptomycin resistance were first described by KITAMOTO et al. (1956).

ROSENKRANZ (1964) pointed out that the mode of resistance in strains bearing R-factors is different from that found with chromosomal mutants. He proved that ribosomes extracted from an R^+ strain of *E. coli* are inhibited by the drug in a cell-free protein-synthesizing system, as well as ribosomes extracted from the parent sensitive strain. These results favour the hypothesis that the streptomycin resistance of R^+ cells is located at a different site from the ribosomes, and it was thought that it could be due to modifications of cell permeability.

Soon after, OKAMOTO and SUZUKI (1965) demonstrated the presence of enzymes able to inactivate kanamycin and streptomycin in a strain of *E. coli* carrying an R-factor. The sensitive parent strain was devoid of any enzymatic activity. An absolute requirement for acetate as acetyl-coenzyme A led them to assume that kanamycin inactivation was due to acetylation of the drug on an undetermined part of the molecule. For streptomycin the exact reaction could not be precisely determined.

The first detailed observations on the inactivation process of some aminoglucosides by R^+ strains were described by UMEZAWA et al. (1967). This work confirmed the data of OKAMOTO and SUZUKI, showing that, with the R-factor studied, the amino group of 6-amino-6-deoxy-D-glucose of kanamycin is acetylated. The correlation between inactivation and resistance is obvious since the bacterial cells from which the enzyme is extracted are sensitive to kanamycin C, which contains a molecule of D-glucosamine instead of the 6-amino-6-deoxy-D-glucose; moreover, the enzyme does not inactivate kanamycin C.

UMEZAWA et al. (1967) also described a mode of inactivation of kanamycin, dihydrostreptomycin and paromomycin by an ATP-dependent enzymatic system. In this case they studied another R-factor carrying resistances for these three antibiotics. With the observation that, by treatment with alkaline phosphatase, the activity of the drug could be recovered (about 60% for kanamycin and 100% for dihydrostreptomycin) the authors concluded that phosphorylation was responsible for inactivation.

These results were summarized by KONDO et al. (1968) as shown in Table. 9.

Streptomycin inactivation by enzymatic processes controlled by extrachromosomal elements is presented by YAMADA et al. (1968). The R-factor studied was isolated in a wild-type strain of *E. coli* and the enzyme was obtained by the osmotic shock method of NOSSAL and HEPPEL (1966). The inactivation of streptomycin is demonstrated by a biological or biochemical assay to detect the formation of a mono-adenylphosphorylated derivative of the drug, using ATP-α-^{32}P as the adenylation donor.

The enzyme seems to be an adenyl-synthetase, produced by the cells under the control of the plasmid. It is formed in equal amounts with or without

Table 9. Inactivation of aminoglucosidic antibiotics by resistant microorganisms

Antibiotic	Bacterial strains		
	E. coli K12 (R5) R-factor isolated in a *Shigella*	*E. coli* K12 (MC1629) R-factor isolated in an *Escherichia*	*Pseudomonas* naturally resistant strain
Kanamycin	Acetylation of amino group of 6-amino-glucose	Phosphorylation of hydroxyl group in position C_3 of 6-amino-glucose	Phosphorylation of hydroxyl group in position C_3 of 6-amino-glucose
Paromomycin	—	Phosphorylation of hydroxyl group in position C_3 of 2-amino-glucose	Phosphorylation of hydroxyl group in position C_3 of 2-amino-glucose
Streptomycin	—	Unlocalized adenyl-ation[a]	—

[a] The enzyme of YAMADA *et al.* (1968) leads to an adenylation on the 3-hydroxyl group of the N-methyl-L-glucosamine moiety.

the drug. However, it has an absolute requirement, as substrate, for streptomycin or a very closely related compound. The other aminoglucosides (neomycin, kanamycin, paromomycin and gentamycin) are not inactivated by this enzyme. The only requirements for the modification of streptomycin are ATP, Mg^{++} ions and a monovalent cation, such as NH_4^+. The inactivated product is an o-adenylated streptomycin in which an o-adenylye group is attached to the 3- hydroxy group of the N- methyl-L-glucosamine. The adenylated derivative is totally inactive as an antibacterial substance. It does not promote growth of streptomycin-dependent mutants and if used in an *in vitro* protein-synthesizing system, directed by poly-U with ribosomes extracted from a sensitive strain, it provokes no miscoding and no inhibition of the polypeptide synthesis.

HARWOOD and SMITH (1969) obtained similar results when they studied another strain of *E. coli* carrying a multiple-resistant R-factor with resistance to streptomycin, tetracycline, chloramphenicol and sulphonamides. They showed that the ribosomes of this strain are normally sensitive to streptomycin. They concluded that the adenyl-synthetase, like a plasmid-controlled penicillinase, is a periplasmic enzyme, easily released by osmotic shock.

As this transferable resistance character is dominant over the sensitive one, it would be interesting to determine to what extent this mode of resistance is found in wild-type strains of enterobacteria resistant to streptomycin. Moreover, it has been noticed that strains of *E. coli* or *Salmonella* resistant to new aminoglucosides still rarely used in therapeutics (such as bluensomycin, spectinomycin or gentamycin) are isolated more and more frequently. These resistances are due to R-factors, bluensomycin resistance being normally associated with streptomycin resistance and resistance to other drugs controlled by new determinants. The mechanisms of resistance to these new antibiotics are not fully understood, but they do not seem to depend upon modifications at the ribosomal level, which is the precise site of action of these drugs. With *in vitro* protein-synthesizing systems, these antibiotics inhibit protein synthesis at the same level, whether the ribosomes are extracted from sensitive or from resistant strains (SMITH, 1967).

8. Streptomycin Resistance in Some Strains of *Enterobacteriaceae*

In order to determine to what extent the enzymatic inactivation process is involved in streptomycin resistance, several strains of wild-type *E. coli* and *Salmonellae* were studied (RASSEKH and PITTON, 1971).

After determination of the MIC for streptomycin, using different culture media at different pH values and with various inocula, transferability of the resistance character was tested according to the method of ANDERSON and LEWIS (1965a): the sensitive recipient strains were *S. typhi*, *S. paratyphi* B and *S. typhimurium* for the wild-type strains of *E. coli* as donor strains and *E. coli* K12F^{+} and K12F^{-} when the donor strains were the wild *Salmonellae*. 54 strains of *E. coli* were used; for 50, i.e. 92%, streptomycin resistance was transferable. For 20 strains of *Salmonella*, 10 strains, i.e. 50%, were able to transfer their resistance.

The inactivation tests, which were done with whole cells—by adding streptomycin to a growing culture and determining the residual antibacterial activity after incubation—were inconclusive. The residual activity was high, about 70% of the initial amount of the drug; moreover, only 7 out of the 74 strains tested were able to carry out this partial inactivation.

In contrast to this, the cell-free tests done with proteins obtained either by osmotic shock or after extraction of total proteins by grinding have shown that the majority of the wild-type strains studied contains inactivating enzymes for streptomycin. Complete inactivation occurs normally in 4 to 6 hours at 37° C and is dependent upon the presence of ATP and Mg^{++}ions.

In the *in vitro* system, streptomycin was inactivated by all the 54 wild-type strains of resistant *E. coli*. This inactivation occurred also with the strains which could not transfer their resistance. The results obtained strongly support the hypothesis that streptomycin resistance among wild-type strains of *E. coli* is probably always due to R-factors. The fact that some of these strains did not transfer their resistance does not necessarily mean that their resistance is due to a mutation on the chromosome. Most probably they are R-determinants devoid of transfer factor, according to the model of ANDERSON.

The result of inactivation tests for wild-type strains of *Salmonellae* showed that 60% of these strains inactivate the antibiotic. It is important to point out that none of the strains which were unable to inactivate the drug transferred their streptomycin resistance. It is thus possible that resistance is controlled by a mutation in a chromosomal gene resulting in an altered ribosomal protein.

The same bacterial strains which inactivated streptomycin in the *in vitro* system showed very weak inactivation properties in the *in vivo* system. It is known that the adenylated or phosphorylated derivatives of the drug are without action on protein synthesis, that is to say, they are not able to introduce miscoding as streptomycin does. It can then be supposed that, after enzymatic inactivation, these derivatives are still capable of binding to ribosomes but without miscoding activity. In that case, they could exert a feedback inhibition on an active transport system of the drug into the bacterial cell. It is also possible that these derivatives play a major role in maintaining osmotic equilibrium, their intracellular presence inhibiting, by physico-chemical mechanisms, penetration of the drug into the cytoplasm.

VI. Chloramphenicol

This antibiotic, in addition to its broad spectrum of antibacterial activity, damages bone-marrow cells. It has stimulated a great deal of work on its mode of action as well as on the mechanism of resistance (VAZQUEZ, 1966c; HAHN, 1967). This work is important for the understanding of the relationships existing at the ribosomal level between the different components or subunits of ribosomes and substances active in the inhibition of protein synthesis.

Chloramphenicol inhibits the growth of the majority of Gram-positive and Gram-negative bacteria, but has no action on yeasts and fungi, although mitochondria of yeasts and certain mammalian cells are known to be sensitive to the antibiotic (KROON, 1965).

Chloramphenicol is normally absorbed by sensitive bacterial cells but seems not to be firmly bound; it can be recovered by repeated washings. This phenomenon is consistent with the reversibility of its bacteriostatic activity. Absorption is dependent upon incubation conditions and the physiological state of the bacterial cell, the uptake rate being initially very high but falling rapidly after a few minutes.

A. Mode of Action

In a bacterial culture, the addition of chloramphenicol in a growth-inhibiting concentration results in inhibition of protein synthesis, although the energy-yielding processes are generally unaffected (CHUIT, 1968). The antibiotic promotes an increase in the total content of nucleic acids but has no influence on synthesis of the cell wall.

1. Site of Action

Chloramphenicol, a typical bacteriostatic drug, inhibits protein synthesis at the ribosomal level. In cell-free systems, the antibiotic binds specifically to 50S subunits and normally not to the 30S or other cellular components. The maximum binding in appropriate conditions seems to be about one molecule for 3.4 ribosomes (VAZQUEZ, 1964b, 1966a, 1966b). This result, however, is not conclusive, since it is likely that a proportion of ribosomes is inactive after extraction and purification. It can be supposed that in intact cells there is at least one binding site for chloramphenicol on each ribosome (WOLFE and HAHN, 1965), although it is possible that there may be more drug-binding sites per ribosome unit.

2. Influence on Nucleic Acid Synthesis

GALE and FOLKES (1953) first showed that, when staphylococci were grown in presence of chloramphenicol, increased synthesis of nucleic acids could be demonstrated, especially in the RNA. DNA synthesis was only slightly modified. In cell-free systems, DNA and RNA polymerases are not affected by chloramphenicol (ELLIOTT, 1963; WARING, 1965). The structure of the accumulated RNA is not determined, but it seems likely that it is principally t-RNA (KURLAND et al., 1962; NOMURA and HOSOKAWA, 1965) and m-RNA (HAHN and WOLFE, 1961). It has been suggested that the RNA accumulated in the treated cells could either be a repressor substance or operate as an m-RNA coding for synthesis of a special protein acting as a repressor on protein synthesis (SYPHERD and STRAUSS, 1963; BOWNE and ROGERS, 1963).

3. Effect on Protein Synthesis

Following its binding to bacterial ribosome, chloramphenicol could inhibit protein synthesis at one of the following steps:

a) Formation of the m-RNA-ribosome-aminoacyl-t-RNA complex.

b) Peptide elongation.

c) Release of nascent peptide chain.

Since t-RNA and the chloramphenicol bind on the same ribosomal subunit, one might have thought the drug inhibited the binding of t-RNA or aminoacyl-t-RNA on its specific site. The results of CANNON et al. (1963) have shown that this is not the case, and it is now known that in cell-free systems chloramphenicol has no significant activity on the binding of phenylalanyl-t-RNA to ribosomes when the system is directed by poly-U (NAKAMOTO et al., 1963; WOLFE and HAHN, 1965).

Although the antibiotic binds on 50S subunits and m-RNA on 30S subunits, it has been supposed that the drug could interfere with messenger binding by an allosteric effect. SPEYER et al. (1963) have shown that chloramphenicol inhibits the incorporation of proline in a system directed by poly-C, the incorporation of lysine directed by poly-A, but does not significantly inhibit phenylalanine incorporation directed by poly-U.

KUCAN and LIPMANN (1964) pointed out later that amino-acids incorporation directed by native m-RNA bound to ribosomes is less sensitive to chloramphenicol than incorporation directed by a messenger prepared from the f2 phage of *E. coli*.

These results support the hypothesis that the binding of m-RNA on ribosomes is inhibited by the drug. Some authors consider that the antibiotic competes with m-RNA for some specific ribosomal binding sites (WEISBERGER et al., 1964).

Finally, the hypothesis has been developed that chloramphenicol stops protein synthesis by inhibiting polypeptide chain elongation. JULIAN (1965) showed that, whereas the formation of di- or tri-peptides is not affected, synthesis of longer chains is strongly inhibited. TRAUT and MONRO (1964) established that the antibiotic counteracts the effect of puromycin by the release of poly-phenylalanyl-t-RNA elements bound to ribosomes. They suggested that chloramphenicol acts, like puromycin, as an inhibitor of the enzyme responsible for the formation of peptide bonds. This has been confirmed by the results of MONRO and MARCKER (1967); MONRO (1967) and DAS et al. (1966).

This information is now interpreted in terms of an inhibitory effect of chloramphenicol on the peptidyl-transferase, which is a structural component of the 50S ribosomal subunit and which forms the peptide bond.

B. Chloramphenicol Resistance

Summarizing the literature on this problem, one may conclude that the inability of chloramphenicol to inhibit growth of microorganisms can be due to one of the following mechanisms:

a) Reduced permeability of the cell membrane; this has been demonstrated for resistant mutants of *E. coli* (VAZQUEZ, 1964a).

b) Inability of the drug to bind to ribosomes (VAZQUEZ, 1964a); this is certainly true for cells of yeasts and protozoa with 80S ribosomes.

c) Destruction or inactivation of the drug (MERKEL and STEERS, 1953; MIYAMURA, 1964).

d) Impairment of protein-synthesizing processes (RAMSEY, 1958).

Probably only a) and c) should be considered at the present time for pathogenic bacteria.

1. Chromosomal Control of Chloramphenicol Resistance

Location of genes. With Hfr crosses between different strains of *E. coli* K12, CAVALLI and MACCACARO (1952) have shown that chloramphenicol resistance may have a mutational origin; a cluster of genes located between the loci *thi* and *met* at the end of the chromosome of an Hfr Hayes strain could confer resistance to both chloramphenicol and tetracycline.

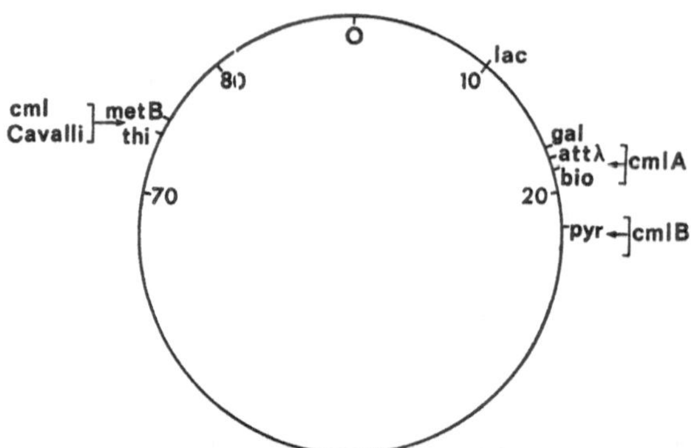

Fig. 9. Chromosomal location of genes responsible for resistance to chloramphenicol. The figure shows the approximate location of the genes controlling chromosomal resistance to chloramphenicol. According to REEVE (1968), *cmlA* is located near *attλ* and *cmlB* near *pyr*; according to CAVALLI and MACCACARO (1952), these genes are close to *metB* and *thi*

More detailed information has been produced by REEVE (1968), REEVE and DOHERTY (1968) and REEVE and SUTTIE (1968) on the mapping of the genes responsible for chloramphenicol resistance. The data of REEVE lead to the following conclusions: mutations conferring various levels of resistance to chloramphenicol (and to tetracycline) can be detected after plating *E. coli* suspensions on low concentrations of chloramphenicol. These mutations can occur at three chromosomal loci:

a) One gene, named *cmlA*, is very close to the site of phage *λ* attachment; it confers a high level of resistance to chloramphenicol and a very low level of resistance to tetracycline.

b) One gene, named *cmlB*, located near *pyrD*, gives low resistance to chloramphenicol but high resistance to tetracycline.

c) One gene, not precisely located, gives mutants with low resistance to both antibiotics.

The genetic map of these mutants is shown in Fig. 9. The phenotypic expression of *cmlA* and *cmlB* genes is different. The A locus is easily transduced by phage P1 to sensitive cells, with a very short phenotypic lag. This does not apply to the B locus, which can be transduced only at very low frequency. This occurs only after several cell divisions have occurred after phage treatment and before plating on selective media containing the antibiotic so that the phenotypic gene expression can take place..

2. Extrachromosomal Control

Chloramphenicol resistance mediated by transferable R-factors has been described by Japanese workers (WATANABE. 1963),

A striking feature is that the determinants for chloramphenicol resistance are always carried by *fi*+ transfer factors. This suggests that homology exists between the base sequence of these factors and the sequence of the corresponding determinants. Moreover, the association between the R-determinant and the transfer factor must be close since it is normally difficult to separate the two units, even with interrupted mating mixtures.

C. Mechanisms of Resistance to Chloramphenicol

The early studies could not distinguish resistance due to chromosomal genes from that mediated by extrachromosomal elements. MERKEL and STEERS (1953) first suspected the existence of enzymic inactivation of the drug. These authors thought that a chloramphenicol-reductase was responsible for resistance, whereas KUSCHNER (1955) considered that permeability of the cell could be modified, thus inhibiting the uptake of the drug.

1. Modifications of Cell Permeability

Studying amino-acid incorporation, OKAMOTO and MIZUNO (1962) found that cell-free extracts prepared from resistant *E. coli* B are as sensitive to the drug as extracts prepared from the sensitive parent strain. Other data have led OKAMOTO and MIZUNO (1964) to suppose that, since resistant strains lose their resistance to chloramphenicol in cell-free extracts, the primary reason for insensitivity was not an alteration in the protein-synthesizing system, but a modification of one step of the process by which the compound reaches its site of action. Because their resistant cells showed cross-resistance with tetracycline, they supposed that resistance could be due to a reduced cell permeability. These results have been obtained with resistant mutants as well as with strains carrying R-factors.

The data of UNOWSKY and RACHMEIER (1966) are based on a study of R^+ strains only. These authors develop a rather complicated theory in which the *fi* character of the transfer factors could play a part in the permeability phenomena: cells carrying fi^- factors with determinants for tetracycline and streptomycin resistance would be impermeable to both antibiotics, whereas cells carrying fi^+ factors with resistance to chloramphenicol, tetracycline and streptomycin would be freely permeable to streptomycin and tetracycline but impermeable to chloramphenicol.

2. Enzymic Mechanisms Implicated in Chloramphenicol Inactivation

MERKEL and STEERS (1953) and MIYAMURA (1964) called attention to the importance of enzymic inactivation in resistance to chloramphenicol. The latter pointed out that a very high proportion of the chloramphenicol-resistant strains of *Shigella*, *Escherichia* and *Staphylococci* isolated from clinical specimens can degrade the drug when grown in presence of chloramphenicol. The 29 resistant strains of *Pseudomonas* studied did not show any inactivation of the drug. Out of 235 sensitive strains, none was able to produce inactivation. Some resistant mutants isolated *in vitro* from sensitive cells sometimes show a low degree of inactivation.

OKAMOTO and SUZUKI (1965) first described a strain of *E. coli* carrying an R-factor isolated in *Shigella* which was able to inactivate chloramphenicol in cell-free systems containing acetylcoenzyme-A as an acetyl donor. They assumed from these experiments that the antibiotic was inactivated by acetylation.

BOUANCHAUD (1967) tried to determine the specificity of some resistance characters. For chloramphenicol, he considers that bacterial resistance can be due either to a reduced cell permeability or to an esterification by acetyl group. The colorimetric technique he used demonstrated the presence of esterified derivatives.

The data of OKAMOTO et al. (1967) permitted further insight into the acetylation reactions in Gram-negative bacilli. Working with ^{14}C-chloramphenicol, they studied the esterification capacity of 36 strains of *Enterobacteriaceae* (*Salmonella*, *Shigella*, *Escherichia*, *Klebsiella*, *Aerobacter*, *Serratia*, *Proteus* and *Pseudomonas*). The majority of *Proteus* strains, whether sensitive or resistant, show a significant rate of acetylation. Strains of *Pseudomonas*, although resistant to the drug, revealed only a low acetylation ability, apparently insufficient to explain the high resistance level of the cells; in these strains, reduced permeability of the membrane or a mode of inactivation different from acetylation could be involved. The enzyme found in the *Proteus* strains studied by OKAMOTO yields mono- and di-acetylated derivatives of chloramphenicol, which is a reaction similar to that found with *E. coli* strains carrying R-factors.

It is interesting to note that the acetylase isolated in *S. aureus* differs from that of Gram-negative bacilli, since it does not produce the diacetylated derivative, even with a high degree of mono-acetylation (SUZUKI et al., 1966b).

Finally, the data of SHAW (1966, 1967), SHAW and BRODSKY (1967), MISE and SUZUKI (1968) and WINSHELL and SHAW (1969) have explained most of the biochemical mechanisms concerned with *in vitro* acetylation reactions.

3. Chloramphenicol Acetylation in *E. coli*

The study of cell-free systems reveals that the pathway by which chloramphenicol is inactivated consists in the formation of a mono-acetylated derivative before diacetylation. Chromatographic and spectral analyses indicate that the 3-acetoxy ester is accumulated in this system. Some observations suggest, however, that the mechanism could be more complicated, and the detection of small quantities of the 1-acetoxy derivative raises the question whether this compound is an intermediate in the biosynthesis, a secondary product or a degradation product. The reaction pathway is shown in Fig. 10.

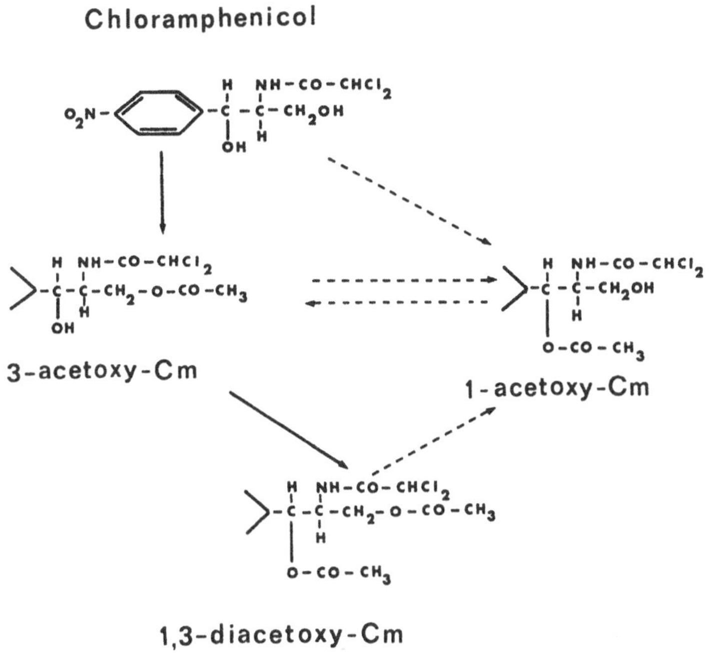

Chloramphenicol

3-acetoxy-Cm

1-acetoxy-Cm

1,3-diacetoxy-Cm

Fig. 10. Pathway of acetylation reactions for chloramphenicol. From chloramphenicol, the first derivative obtained is the 3-acetylated derivative, and from this latter the 1,3-diacetylated one. The question is whether the 1-acetylated derivative is an intermediate in biosynthesis or a secondary product of degradation

It is quite possible that the 1-acetoxy ester is a secondary product, since it cannot be used as a substrate for the addition of a second acetyl radical, whereas 3-acetoxy-chloramphenicol is an excellent substrate to obtain the diacetylated drug. The enzyme would thus be a trans-acetylase or an acetyltransferase. The results of SHAW should show that synthesis of the enzyme is a constitutive property, at least in the *E. coli* strain studied.

4. Production of the Enzyme

The fact that the enzyme is synthesized by various bacterial strains without R-factors, or so it is supposed, led to the development of two hypotheses concerning the mode of enzyme production in the strains bearing such factors:

a) that the resistance gene carried by the R-factor is the structural gene for the synthesis of the acetyltransferase.

b) that the synthesis of the enzyme is directed by chromosomal genes; the plasmidic genes would then be responsible for the synthesis of an intermediate substance capable of derepressing the chromosomal determinant, which is normally under strong repression in the sensitive strain.

The data of MISE and SUZUKI (1968) favour the first hypothesis. The fact that several bacterial strains can grow at 34° C but not at 43° C in presence of chloramphenicol, added to the fact that these strains produce a thermolabile acetyl-transferase, would confirm such an assumption. Moreover, the existence of a thermolabile enzyme in the R^+ temperature-sensitive strains indicates not only that chloramphenicol resistance is really due to the enzymatic activity but also that the synthesis of the enzyme is directly under the control of the extrachromosomal determinant.

If it were possible to prove without any doubt the existence of chromosomal genes responsible for acetyl-transferase synthesis, it would be easier to study the similarity or the differences between enzymes synthesized under the control of chromosomal genes and those synthesized under the control of extrachromosomal genetic material. Such a study could provide important information on the origin of some transferable determinants, permitting finally the solution of the problem of the chromosomal origin of these characters.

Although it is not possible to decide whether one or two enzymes are responsible for the formation of the different acetylated derivatives of chloramphenicol, we may postulate that there is only one enzyme, since the acetylating activity of temperature-sensitive strains shows the same thermolability in each step of the reaction of acetylation.

5. Chloramphenicol Inactivation by Wild-Type Strains of *E. coli*

The data on chloramphenicol acetylation have generally been obtained with cell-free systems, using a relatively small number of resistant strains. It was thought (PIFFARETTI and PITTON, 1970) that an investigation of acetylation with intact cells could contribute to the knowledge on chloramphenicol resistance, from the biochemical as well as from the genetic or epidemiological points of view.

The *in vivo* study of the kinetics of the reaction was able to distinguish the different steps by which the antibiotic is inactivated by resistant bacterial cells.

Several properties of wild-type strains of *E. coli* were determined, especially the following:

a) Transferability of chloramphenicol resistance was tested according to ANDERSON and LEWIS (1965a).

b) MIC determinations were done by serial dilutions in nutrient broth.

c) The assay of residual antibacterial activity (inactivation test) was done by an agar diffusion technique with *S. lutea* as indicator strain (PIGUET and PITTON, 1968).

For the analysis of the formation of metabolites, a thin-layer chromatographic (TLC) test was set up, permitting the detection of about 6×10^{-9} g of the acetylated compounds.

The experiments gave the following results:

Transferability. Out of the 98 wild-type *E. coli* strains, 75 were able to transfer their resistance to the various *Salmonellae* used as recipient strains.

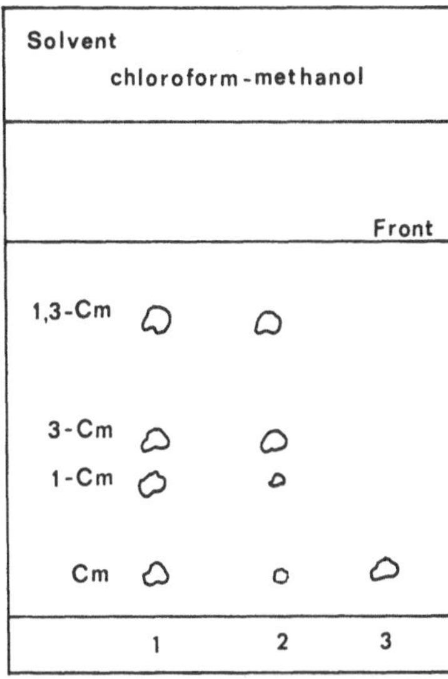

Fig. 11. Thin-layer chromatography of chloramphenicol metabolites. The figure shows the results obtained by thin-layer chromatographic analysis of chloramphenicol metabolites obtained with a wild-type resistant strain and a wild-type sensitive strain of *E. coli*. In position 1, the four spots corresponding to the reference substances can be detected: *Cm* chloramphenicol; *1-Cm* 1-monoacetylated derivative; *3-Cm* 3-monoacetylated derivative; *1,3-Cm* 1,3-diacetylated derivative. In position 2, the metabolites synthesized by the wild-type resistant strain can be seen. In position 3, the result of an experiment done with a wild-type sensitive strain shows only the spot corresponding to unaltered chloramphenicol

Among the others, 15 were sensitive to nalidixic acid and were mated with *E. coli* K12 resistant to this drug; 5 of the 15 were able to transfer their resistance. This gave a total of 84% positive in the transfer tests. Several mating mixtures were done using as donor strains some *E. coli* K12 selected *in vitro* for chloramphenicol resistance; these have always given negative results in transfer tests.

Acetylation and inactivation. The TLC test, a result of which is given in Fig. 11, demonstrated that all the wild-type resistant strains were able to acetylate the drug.

Inactivation tests, carried out simultaneously, showed a proportional decrease of antibacterial activity. The inactivation rate depends upon the reaction medium, the strain under test and the growth phase.

Kinetics of reaction. The experiments on the kinetics of acetylation were carried out with several resistant *E. coli* strains. Some results are shown in Fig. 12 to 15.

In the experiment corresponding to Fig. 12 the drug was added simultaneously with the bacterial inoculum. During the lag phase inactivation is slow, the residual activity being about 85% after 3 hours. At this point, TLC shows that the major part of the antibiotic content is chloramphenicol with a small quantity of the 3-acetoxy derivative. Soon after, a rapid decrease

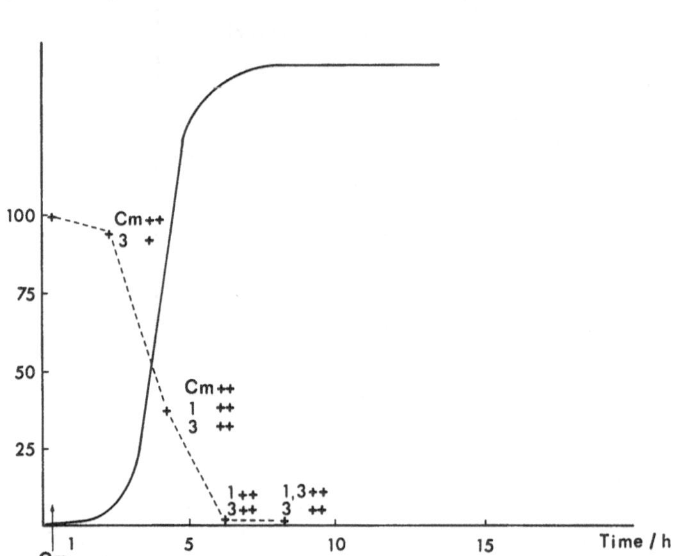

Fig. 12. The antibiotic was added at the beginning of the lag phase

Figs. 12–15. Kinetics of chloramphenicol acetylation in a cellular system. In these figures, the continuous line corresponds to the bacterial growth curve and the dotted line to the rate of chloramphenicol inactivated for a given time. The quantity of the various metabolites is expressed as follows: + weakly coloured spot; ++ large and strongly coloured spot

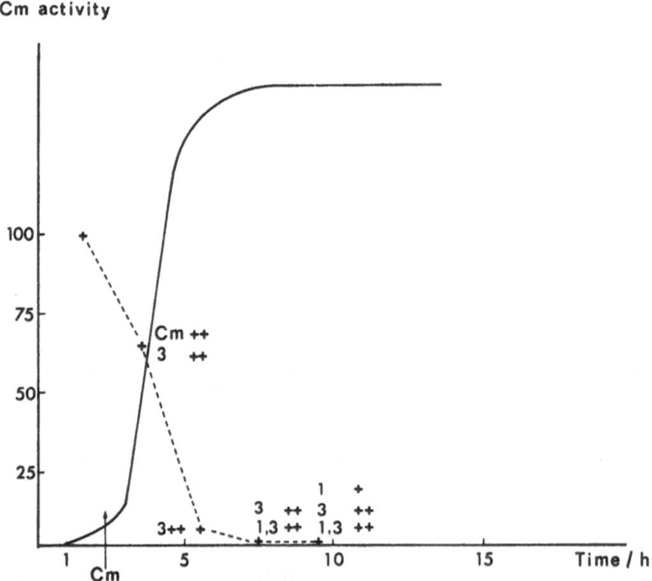

Fig. 13. The antibiotic was added at the beginning of the exponential phase

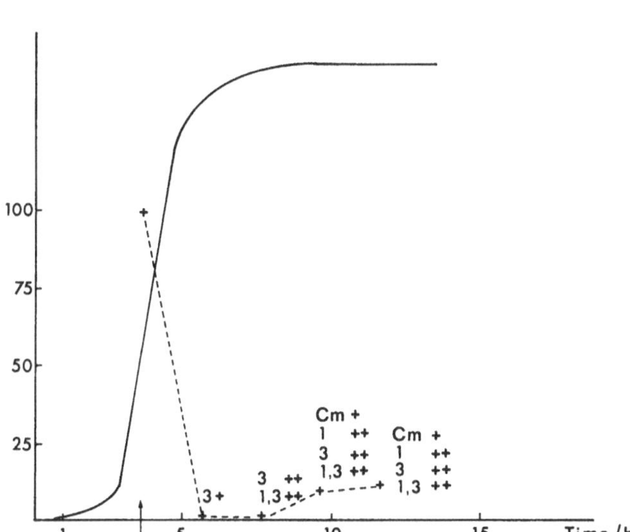

Fig. 14. The antibiotic was added in the middle of the exponential phase

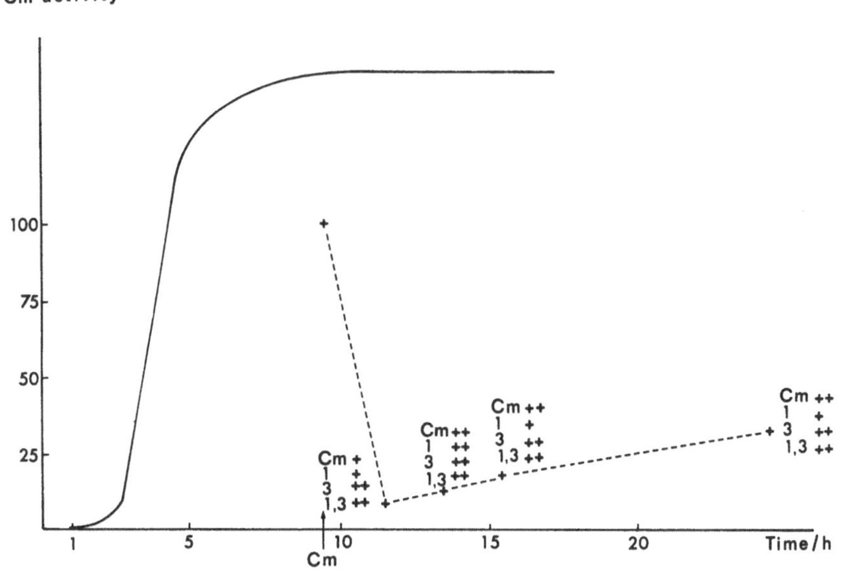

Fig. 15. The antibiotic was added at the beginning of the stationary phase

of activity is detected, with the formation of 1-acetoxy and later 1,3-diacetoxy derivatives. After 7 hours incubation, inactivation is complete.

In the second experiment (Fig. 13), the antibiotic was added at the end of the lag phase; it can be seen that the rate of inactivation is very high. With TLC, both the 3- and the 1,3-diacetoxy derivatives were detected but no chloramphenicol, even after 10 hours' incubation.

The experiments reported in Fig. 14 and 15 show that the addition of the antibiotic during the exponential or stationary phases induces very fast inactivation with formation of the three acetylated compounds. However, after some hours, it can be demonstrated that these derivatives are partially converted into biologically active chloramphenicol. This phenomenon has been confirmed both by biological assays and by TLC tests which show a spot corresponding to the intact drug.

It seems probable that chloramphenicol is acetylated principally during the exponential growth phase; the physiological state of the cells seems to be more important than their number.

Kinetic experiments carried out in buffer solutions, Tris or phosphate, have demonstrated that in the inactivation tests with intact cells, the energy sources (such as glucose) are more impotant than the presence of exogenous acetyl-coenzyme A.

None of the tests done with wild-type sensitive strains of *E. coli* or with different variants of *E. coli* K12 enabled the detection of acetylated chloramphenicol.

The fact that 16% of wild-type strains of *E. coli* do not transfer their resistance does not imply that this character is due to a chromosomal mutation. The transfer factor can be defective or the R-determinant can be integrated on the chromosome, although the experiments of PIFFARETTI (personal communication) with numerous Hfr crosses do not lend any support to this hypothesis. Restriction phenomena may be involved, or the transfer of the resistance character may be inhibited by mechanisms which depend upon the specificity of the strains. However, it seems better to adopt Anderson's model, and to accept that strains in which the resistance determinant is not transferable are devoid of transfer factor and carry only an R-determinant coding for synthesis of the acetyl-transferase.

PIFFARETTI et al. (1970) have further investigated the mechanism of chloramphenicol resistance and have compared the effect of the acetylated compounds *in vivo* on bacterial growth and *in vitro* on protein synthesis with viral RNA as messenger. It was found that chloramphenicol and its acetylated derivatives show the same pattern of inhibition *in vitro* and *in vivo*. The in-

Table 10. The minimum inhibitory concentrations of chloramphenicol and its acetylated compounds (μg/ml)

	E. coli K12	*E. coli* wild-type
Chloramphenicol	3–6	6–12
3-acetoxy-chloramphenicol	50–100	50–100
1,3-diacetoxy-chloramphenicol	200	200

The sensitive strains used are: *E. coli* K12 F^+ met^- azi^s str^s and *E. coli* R376 wild-type (clinical isolate).

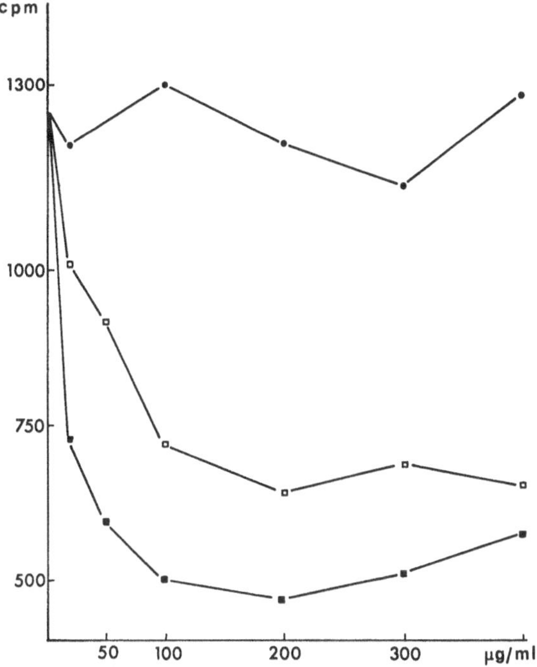

Fig. 16. Incorporation of labeled amino acids in the presence of chloramphenicol or of its acetylated derivatives. ■—■ chloramphenicol, □—□ 3-acetoxy-chloramphenicol, •—• 1,3-diacetoxy-chloramphenicol

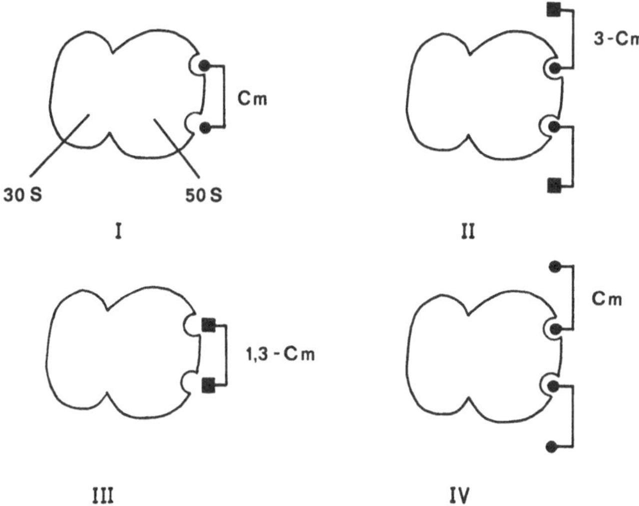

Fig. 17. Proposed model for the binding of chloramphenicol and its acetylated derivatives on ribosomes. In *I*, one chloramphenicol molecule is bound by both its hydroxyl groups; in *II*, two molecules of the mono-acetylated derivative are bound by only one of their OH groups; *III* shows how 1,3-diacetylated-chloramphenicol cannot bind to the 50S subunit and *IV* indicates the reaction which could happen in excess of the drug

hibition decreases with the mono-acetylated compound and almost disappears with the di-acetylated derivative. As shown in Table 10, the MIC differs significantly among the three compounds and Fig. 16 demonstrates the activity of chloramphenicol and the acetylated derivatives on protein synthesis directed by phage R17 RNA as messenger. According to Piffaretti, it is possible to construct a model in which two specific sites on the chloramphenicol molecule correspond to two other sites in the protein-synthesizing system, probably located on the 50S ribosomal subunit. In this model, both sites of the same drug molecule should be attached to the other corresponding sites in order to be completely active. If one of the sites is free, the activity of the antibiotic is decreased. This could explain the partial activity detected *in vivo* and *in vitro* with the mono-acetylated compound. In excess of chloramphenicol, the two sites of the protein-synthesizing system would be occupied by two different molecules of the drug and consequently the antibiotic activity would be decreased. The proposed model of Piffaretti is shown in Fig. 17.

6. The Acetylation in *S. aureus*

The mechanisms of inactivation of chloramphenicol by staphylococci have been explained by the work of Dunsmoor et al. (1963), Suzuki et al. (1966b), Shaw and Brodsky (1967) and Winshell and Shaw (1969).

The main features of the system are the following: the enzyme, in contrast to that synthesized by Gram-negative bacilli, is inducible; it appears in the cell-free extracts only after growth of the cells in sub-inhibitory concentrations of chloramphenicol.

An unexpected analogy between staphylococci and the R^+ strains of enterobacteria has been shown by Chabbert et al. (1964), who demonstrated that the resistance character of *S. aureus* can be lost spontaneously or after acridine treatment, suggesting an extrachromosomal location of the resistance genes. These observations were later confirmed by Sabath et al. (1967).

The optimum pH range of 7.8 is the same for the enzyme extracted from *S. aureus* or from R^+ strains of *E. coli*. Differences are detected in their thermostability. The preparations obtained from *E. coli* are rapidly inactivated by heating at 75° C, whereas those extracted from *S. aureus* are fairly stable at that temperature.

The molecular weight of the enzyme, whatever its origin, has been estimated at about 78.000, determined on sucrose density gradients.

The analysis by disc-gel electrophoresis of both enzymes shows a certain degree of similarity, the electrophoretic mobility being approximately the same. In contrast, the immuno-chemical analyses in double-diffusion tests have revealed basic differences. The antiserum prepared by injection of *E. coli* enzyme gives precipitation lines only with the enzyme extracted from the same cells, but no reaction with the *S. aureus* enzyme. Moreover, the same antiserum neutralizes the acetylating activity of the homologous enzyme, but is without any influence on the staphylococcal enzyme.

MITSUHASHI (1966) summarized the epidemiological and genetic data, giving for chloramphenicol the following conclusions:

a) All the chloramphenicol-resistant strains of staphylococci are able to inactivate the drug.

b) In these strains, the resistance determinant is an extrachromosomal element, but it is not cotransduced with other resistance genes, such as penicillin or erythromycin resistance.

3-Deoxychloramphenicol has a powerful induction property without any antibacterial activity and is not a substrate for the enzyme. This has allowed WINSHELL and SHAW (1969) to specify the induction mechanisms of enzyme synthesis in *S. aureus*. The high resistance levels for chloramphenicol found in the enterobacteria, with MIC higher than 200 μg/ml in contrast to that of staphylococci with MIC about 50 μg/ml, are probably obtained to the detriment of energy production, since the enzyme is used only on presence of the drug. It must be pointed out that no substrate other than the antibiotic or a very close derivative can act as a receptor of the acetyl group under the influence of one or the other enzyme.

Contrary to the success obtained with penicillinase-resistant semi-synthetic penicillins, no therapeutically useful chloramphenicol analogue, which would be neither a substrate for acetylation nor an inducer of enzyme synthesis has yet been discovered. The investigations are still unsuccessful since derivatives which cannot be acetylated are devoid of antibacterial activity (SHAW and BRODSKY, 1968).

VII. Tetracyclines

Since the discovery of this group of antibiotics in 1948, a great deal of work has been undertaken in an attempt to elucidate their mode of action as well as the specific mode of bacterial resistance. The numerous experiments carried out on intact cells, in particular, often show contradictory results.

A. Mode of Action

Three important results can be considered after the exposure of bacterial cells to tetracyclines:

a) Modifications of oxidative and energetic reactions.

b) Influences on nucleic acid synthesis.

c) Influences on protein synthesis.

1. Modifications in Oxidative or Energetic Reactions

In 1950, LOOMIS (1950) showed that chlortetracycline could specifically inhibit phosphorylation.

BRODY et al. (1954) pointed out that the addition of an excess of Mg^{++} ions could reverse the effect on phosphorylation and concluded that the chelating effect of tetracyclines on bivalent ions was responsible for the inhibition.

With the demonstration (SAZ and SLIE, 1954) that nitroreductase synthesis is inhibited by tetracyclines, it was thought that inhibition by tetracyclines was due to chelation of the Mn^{++} ions necessary for the formation of NAD-H_2,which is essential to the activity of nitro-reductase.

SAZ and MARTINEZ (1958) subsequently suggested that the antibiotic exerts its effect by inhibiting electron transport, inducing important disturbances in the energy production systems. Other arguments supporting this hypothesis have been put forward by JONES and MORRISON (1962) after experiments carried out on *K. aerogenes* grown in presence of various tetracyclines.

BENBOUGH and MORRISON (1956) suggested that chlorinated tetracyclines could inhibit bacterial growth by acting on D- glutamate accumulation during growth in aerobic conditions, whereas KRCMERY and KELLEN (1966) thought that oxytetracycline inhibited cytochrome-oxidase and interfered with enzymatic reactions involved in electron transport.

It can be added that it has been shown that tetracyclines can modify oxidation of various compounds active in metabolic processes, such as:

mono-, di-, and tricarboxylic acids,
amino acids: phenylalanine, tyrosine,
sugars: glucose, fructose, xylose,
pyruvate, succinate, acetate and lactate.

Other enzymatic systems that could be influenced by tetracyclines include tryptophanase, penicillinase and β-galactosidase synthesis.

2. Modifications in Nucleic Acid Synthesis

The effect of tetracyclines is very similar to that of chloramphenicol, since nucleic acid synthesis continues during inhibition of protein synthesis. Very little information is available on the type of nucleic acid concerned, but it seems that it is essentially RNA.

YEE and GEZON (1963) have stated that the addition of chlortetracycline to cultures of *Sh. flexneri* could under certain conditions stimulate the production of total RNA.

HOLMES and WILD (1966), investigating the nature of the RNA synthesized by *E. coli* in presence of chlortetracycline, found that 70% is ribosomal RNA, the remainder corresponding to t-RNA.

3. Influence on Protein Synthesis

Although some effects of tetracyclines could be due to their chelating properties on bivalent cations, the fact that all these compounds strongly inhibit *in vitro* protein synthesis has stimulated a great deal of work on their mode of action (GALE and FOLKES, 1953; RENDI and OCHOA, 1962; FRANKLIN, 1963; HIEROWSKI, 1965; GOTTESMAN, 1967; VAZQUEZ and MONRO, 1967). The data obtained by these investigations permit the following conclusions:

a) The concentration necessary to obtain the *in vitro* inhibition, in cell-free systems, is about 10 μg/ml. It is very close to the MIC for cell growth.

b) There is no significant difference in the effects of the various tetracyclines or their related compounds.

c) Chlortetracycline inhibits the incorporation of leucine in ribosomal proteins of *E. coli*, but is does not influence the formation of the amino-acyl-t-RNA complex. The experiments of FRANKLIN (1963) tend to demonstrate that the antibiotic reduces the transfer of the amino acid from the amino-acyl-t-RNA on the growing polypeptide. The addition of excess of Mg^{++} does not prevent inhibition, showing that chelating Mg^{++} is not an important action of the drug.

All tetracyclines reduce the binding of N-acetyl-phenylalanyl-t-RNA in a poly-U-ribosomes system to about 50% (SUAREZ and NATHANS, 1965). In contrast to what is found with chloramphenicol, tetracycline is a powerful inhibitor of polypeptide synthesis (poly-phenyl-alavine) directed by poly-U (LASKIN and CHAN, 1964; PITTON and ALLET, unpublished results).

Other studies have shown that tetracyclines inhibit the binding of amino-acyl-t-RNA to the m-RNA-30S-subunits complexes (SUZUKI et al., 1966a). This suggests that the drug may hinder the binding of t-RNA on the specific acceptor site of amino acids on the 30S ribosomal subunits. More convincing demonstrations of this process have been provided by GOTTESMAN (1967), LUCAS-LENARD and HAENNI (1968) and SARKAR and THACH (1968), who established that tetracycline is able to inhibit the binding of lysyl-t-RNA or of F-met-t-RNA on the specific ribosomal site, the A-site.

Unlike chloramphenicol, tetracycline has no effect on the peptidyl-transferase activity on the ribosomes (TRAUT and MONRO, 1964).

4. Site of Action

Although it is clear that tetracyclines act at the ribosomal level, the studies on the binding of the drug on ribosomal subunits or on other components taking part in the mechanisms of protein synthesis are not conclusive.

The data favouring the hypothesis of a binding on 30S subunits as the specific site of action are as follows:

a) The drug inhibits the binding of aminoacyl-t-RNA on isolated 30S subunits (SUZUKI et al., 1966) but has no effect on the puromycin-dependent release of the peptides (CUNDLIFFE and MC QUILLEN, 1967).

b) When radioactive tetracycline (^{3}H) is added to ribosomes, the antibiotic binds on 50S as well as on 30S subunits (DAY, 1966a); however, this evidence is weakened by the results of CONNAMACHER and MANDEL (1968), LAST et al. (1965) and PITTON and ALLET (unpublished results).

MAXWELL (1968) has shown by equilibrium-dialysis experiments that the binding site must be on the 30S subunits, since the amount of tetracycline bound by the 30S is about double that found on the 50S.

The experiments carried out on the binding to ribosomes are complicated by the fact that the drug can bind strongly to m-RNA and other components of the cell-free systems containing RNA (CONNAMACHER and MANDEL, 1968). It seems that only the fraction bound to ribosomes has an inhibitory activity (DAY, 1966b). The binding depends upon the concentration of Mg^{++} and K^+, and it is still not known whether this binding is completely irreversible, since the inhibitory effects of the drug on intact cells can be suppressed by repeated washings.

B. Resistance to Tetracyclines

The mechanisms responsible for tetracycline resistance of bacterial cells are still not well understood, in spite of the abundant literature on the subject.

1. Chromosomal Control of Resistance

Mapping of the genes. The genetic control of this resistance has also been extensively studied. Most of our knowledge on the location of the genes responsible for tetracycline resistance is due to the work of REEVE (1968). These results show that a close relationship exists between chloramphenicol, puromycin and tetracycline resistance. In addition to the mutants *cmlA* and *cmlB* already described, which were obtained after selection on a medium containing chloramphenicol, REEVE isolated, by selection on tetracycline, three other types of mutants:

a) The first type, mediating low resistance to tetracycline and chloramphenicol has not been precisely located.

b) The second type of mutant, mucoid, promotes low level resistance to chloramphenicol, puromycin and tetracycline; it is probably due to a mutation located near the *capR* locus, the regulator gene for the synthesis of capsular polysaccharides (MARKOWITZ and BAKER, 1967).

c) The third type of mutant is also mucoid, showing increased resistance to tetracycline, no modification in resistance to chloramphenicol and a decrease in puromycin-resistance level. The locus responsible for this type of mutant, although not accurately located, seems to be distant from the *capR* gene.

Thus, these five types of mutants show very different patterns when characterized in relation to their resistance to tetracycline, chloramphenicol and puromycin.

2. Extrachromosomal Control of Resistance

As for chloramphenicol and streptomycin, it is known through the work of the Japanese (WATANABE, 1963) that tetracycline resistance in enterobacteria can be mediated by transferable extrachromosomal determinants.

The data of MAY et al. (1964), ASHESHOV (1966b) and POSTON (1966), and latter those of KASATIYA and BALDWIN (1967) demonstrated that the genes responsible for tetracycline resistance are carried on a plasmid in several strains of *S. aureus*. It is known that the strains in which penicillinase synthesis is due to an extrachromosomal gene can lose this character at relatively high frequencies. The rate of emergence of tetracycline-sensitive variants observed with numerous staphylococci is comparable with the frequency found for determinants responsible for penicillin resistance. Moreover, the loss of these determinants is irreversible.

C. Mechanisms of Resistance to Tetracyclines

Among the major antibiotics, this group is certainly the most complicated and the least well understood with respect to the mechanism of resistance in the various bacterial species. In order to clarify the problem, it is not necessary to take into consideration the relation between the mode of resistance and the mode of genetic control, since the majority of data have not brought to light any obvious differences between the chromosomal and the plasmidic type of control.

Three different types of mechanism can be involved in tetracycline resistance:

a) Inactivation of the drug,
b) Resistance of a ribosomal subunit,
c) Reduced permeability of the cell membrane.

1. Destruction or Inactivation of the Drug

Little information is available on the inactivating ability of tetracycline-resistant bacterial strains. TSUKAMOTO et al. (quoted by OKAMOTO and SUZUKI, 1965) have claimed that such inactivation processes can be found with some strains. These results have not been confirmed.

2. Resistance of Ribosomal Subunits

AKIBA and YOKOTA (1962) have obtained from a tetracycline-resistant strain isolated *in vitro* a cell-free extract resistant to the antibiotic; they assumed that the resistance was located on one of the ribosomal subunits. The results of Okamoto and Mizuno (1964) are completely different, as are those of Pitton and Allet (unpublished results) working with various wild-type strains of *E. coli*, with or without transferable tetracycline determinants.

Ribosomal influences on tetracycline resistance. Two points have been studied in this investigation:

a) The activity of tetracycline on protein synthesis, in an *in vitro* cell-free system, using poly-U as messenger and studying the incorporation of ^{14}C-phenylalanine.

b) The possibility of determining, by disc-gel electrophoresis, some differences in the nature or even the composition of ribosomal proteins.

Table 11. Resistance of *E. coli* to tetracyclines. Characteristics of the strains

Designation of the strain	Resistance pattern	MIC for tetracycline	Character of the transfer factor
E. coli K12F$^-$	sensitive	—	
E. coli R102 wild-type	T	140 µg/ml	non-transferable
E. coli R401 wild-type	T	300 µg/ml	*fi*$^-$
K12 F$^-$ (R401)	T	140 µg/ml	T determinant transferred from R401
E. coli R214 wild-type	T	50 µg/ml	*fi*$^+$
K12 F$^-$ (R214)	T	20 µg/ml	T determinant transferred from R214
K12 F$^-$ Tc-r20	T	20 µg/ml	Mutant resistant to tetracycline
E. coli R278 wild-type	T	340 µg/ml	*fi*$^-$
K12 F$^-$(R278)	T	140 µg/ml	T determinant transferred from R278
E. coli R52 wild-type	C	—	
E. coli R47 wild-type	AS	—	
*E. coli*B	sensitive		

The strains used for the preparation of ribosomes are described in Table 11. Some of the R-factors were transferred into *E. coli* B, a laboratory strain for which the composition of ribosomal proteins is well known.

Antibiotic activity in in vitro cell-free extracts. The type of results obtained with all the strains studied is shown in Fig. 18.

It can be seen that whatever the origin of the ribosomes (wild-type resistant strains, wild-type sensitive strains, *E. coli* K12 or *E. coli* B sensitive, or resistant because they carry an R-factor, resistant mutant of *E. coli* K12) the antibiotic exerts its inhibitory activity on polypeptide synthesis. Only

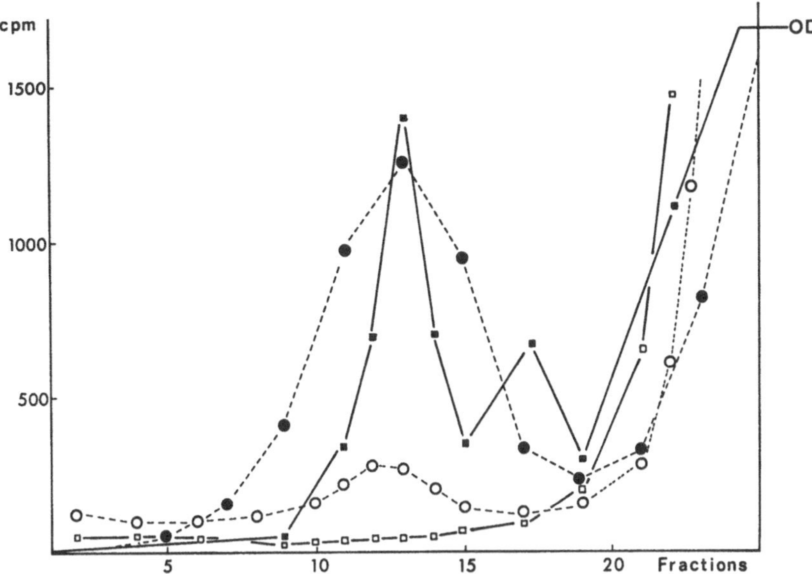

Fig. 18. Incorporation of ^{14}C-phenylalanine and distribution of radioactivity. The ribosomes used in the test were extracted from a wild-type resistant strain of *E. coli*. ▪—▪ Spectrophotometric profile expressed in optical density. •---• Incorporation of ^{14}C-phenylalanine in the control system containing all components except tetracycline. o--o Incorporation of ^{14}C-phenylalanine in the complete system with tetracycline. □—□ Incorporation of ^{14}C-phenylalanine in a system containing tetracycline, but lacking poly-U

some minor quantitative differences can be observed between the different strains.

Ribosomal proteins. The extraction of ribosomal proteins was done by the trichloracetic acid technique. After purification of the 70S ribosomes on a sucrose density gradient containing $MgCl_2$ 10^{-2} M, the ribosomal particles were suspended in an 8 M urea solution. After addition of ribonuclease, the ribosomal proteins were precipitated with trichloracetic acid; the latter was eliminated by several washings with ethyl ether and the proteins were dissolved in an 8 M urea solution. This preparation was used for disc-gel electrophoresis in polyacrylamide gels (7.5% acrylamide in 8 M urea at pH 4.4) according to Traut (1966) and Moore et al (1968).

The electrophoretic pattern of the ribosomal proteins studied is identical for all the strains under investigation, as shown in Fig. 19; there is only one exception, the K line found in all K12 strains, which probably has no importance in the tetracycline resistance mechanism.

In a series of other experiments, the proteins of 50S and 30S subunits were studied for some of the strains described. In this case again, no significant differences have been found in the content of ribosomal proteins, when studied by disc-gel electrophoresis.

These results refute the hypothesis that modifications at the ribosomal level could be responsible for resistance to tetracyclines.

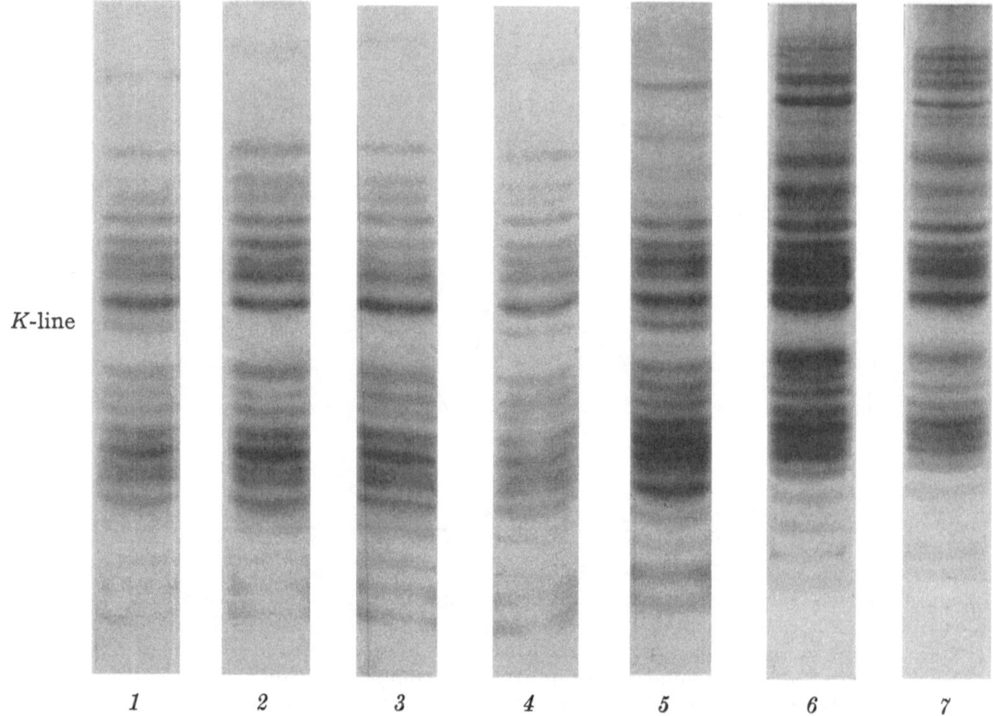

Fig. 19. Electrophoretic pattern of ribosomal proteins from tetracycline-resistant strains. The different gels correspond to the following strains: *1 E. coli* K12 F^- sensitive; *2 E. coli* R102 wild-type strain with a non-transferable resistance; *3 E. coli* R401 wild-type strain with a transferable tetracycline resistance; *4 E. coli* K12 (R401); *5 E. coli* K12 Tc-r20, mutant resistant to tetracycline selected *in vitro*; *6 E. coli* B sensitive; *7 E. coli* B (R401). The arrow shows the position of the *K* line corresponding to a ribosomal protein which is found only in the *E. coli* K12 strains. The anode is on the top of the gels

3. Modifications of Cell Permeability

Many reports have been devoted to this problem. It is therefore necessary to select the most significant ones.

IZAKI and ARIMA (1963) described the accumulation of a large amount of oxytetracycline inside the sensitive cells of *E. coli*, this accumulation being in direct relationship to the antibacterial activity of the drug. This phenomenon seems to depend upon energy-producing systems, since the presence of glucose is necessary for the reaction. IZAKI and ARIMA also demonstrated that in highly resistant bacteria no detectable amounts of intra-cellular oxytetracycline were found. It was not possible to decide from their experiments whether the accumulated substances are bound to cellular components. If the antibiotic binds to some components of sensitive cells, ribosomes for example, one could consider some other mode of resistance than reduced permeability.

But in that case, the drug could not be accumulated inside the cell, since it cannot bind to a resistant component.

The tetracycline-resistant strains isolated by OKAMOTO and MIZUNO (1964) give drug-sensitive cell-free extracts and their explanation for resistance is based on modifications of cell permeability.

FRANKLIN and GODFREY (1965) produced evidence that resistance could be due to a modification of the transport system of the drug through the cell membrane. They consider that a specifically active excretion of tetracycline by resistant cells must be excluded.

IZAKI and ARIMA (1965) studying the accumulation of oxytetracycline by *E. coli* cells, established that this mechanism operates only if the cells are incubated in high concentrations of the antibiotic, and that in the absence of glucose or magnesium sulfate the accumulation is weak.

IZAKI et al. (1966) demonstrated that the accumulation of tetracycline is inversely proportional to the resistance level in multiresistant strains of *E. coli*. However, their results show that the decrease in accumulation capacity is dependent on an induction phase, that is, on the growth of the cells in presence of the drug. But they could not induce this decreased uptake in sensitive cells of the same strain, under the conditions used for resistant cells. As the decrease of accumulation capacity of tetracycline is not modified after the cells have been grown in presence of streptomycin or chloramphenicol, the authors concluded that resistance to the tetracyclines is specifically induced by a member of the group. If this decrease is the real mechanism of resistance, it can be inferred that multiple resistances are not due to one common mechanism but to a system which is absolutely specific for each drug, as is already known for streptomycin, penicillins and chloramphenicol.

UNOWSKI and RACHMEIER (1966) obtained results similar to those of IZAKI and ARIMA (1965). Their study shows that phenomena concerned with cell permeability for tetracycline appear only after the cells have been grown in presence of the antibiotic.

FRANKLIN (1967) considers that some strains of *E. coli* carrying R-factors can have two resistance levels. When the cells are grown for several generations in a medium lacking the drug, and then suddenly transferred into a tetracycline solution, they show a low resistance level. In this case, a concentration, of about 10 μg/ml has little effect on protein synthesis, whereas a concentration of 50 μg/ml brings about a 50% inhibition. After incubation in sub-inhibitory concentrations of tetracycline, the cells reach high resistance levels, and concentrations of 200 μg/ml produce only 45 % inhibition of protein synthesis. The amount of antibiotic absorbed by the cells is 6 to 19 times less than that observed with cells which have been preincubated in media without tetracycline.

De ZEEUW (1968) demonstrated that the accumulation of tetracyclines by *E. coli* cells is a biphasic phenomenon. For concentrations lower than the bacteriostatic level, the uptake must be considered as a simple adsorption on the cell surface. For concentrations higher than the MIC, another mechanism works in addition to the initial adsorption of the drug. This second process, for which energy must be available, probably represents the penetration throughout the cytoplasmic membrane by tetracycline molecules and the transport to their site of action. De Zeeuw believes that tetracycline-resistance is due to a decrease in the active transport of the drug, without any loss of the mechanisms concerned. Probably a higher concentration of antibiotic is required before the accumulation system becomes functional.

Finally, Last et al. (1969) demonstrated the absence of oxytetracycline resistance in cell-free extracts prepared from cells able to grow in drug concentrations 100 times higher than the MIC of the sensitive parent strain.

All this information seems to lead to the conclusion that tetracycline resistance, whatever the mode of genetic control, is due to a permeability barrier; but it is still difficult to determine exactly what mechanism is involved.

VIII. The Epidemiological Aspects of Transferable Drug Resistances

The clinical importance of resistance to antibiotics can vary according to the nature of the genetic control of the resistance. As we have tried to show with the wild-type strains of enterobacteria and staphylococci, the resistances controlled by extrachromosomal determinants are much more important than those controlled by chromosomal genes.

When we are faced with infections involving only a small number of bacteria, we may postulate that the occurrence of resistant mutants is very rare. Their responsibility for the failure of chemotherapeutic treatment is very improbable. The resistance level of these mutants, with the exception of streptomycin-resistant mutants, is often too low to produce real therapeutic resistance. Moreover, since these mutants are often modified in their metabolism, their growth rate is decreased and their pathogenicity is lower than that of the parent strain. Theoretically, combinations of antibiotics could prevent the emergence of such mutants, since the probability of double mutants is very low, although it is well known that in this matter theory is often widely different from reality.

We have seen that the resistances controlled by extrachromosomal elements show a completely different picture. These single or multiple resistances can be transferred alone or "en bloc"; moreover, they promote directly resistance levels which are very dangerous from the therapeutic point of view, since in pathogenicity and virulence these strains are absolutely normal. It must be pointed out that it is possible to accumulate in the same cell several plasmids mediating drug resistances, production of colicins or bacterial toxins.

These bacteria possess an enormous ecological advantage when they emerge in a physiological flora. This advantage is drastically increased when a selective pressure is provided by one, and one only, of the antibiotics included in the resistance spectrum of the cell.

It is obvious that R-factors can spread very widely within the *Enterobacteriaceae* and related bacteria. If one considers *S. typhimurium*, ANDERSON (1968) has shown that in Great Britain the evolution towards transferable resistance was dramatic; in 1961, only 2% of resistant strains could be detected, but the percentage rose to 20% in 1964 and to more than 60% in 1966. These figures represent the high incidence of R-factors in these strains of *Salmonella*. Attention must also be drawn to the increasing frequency of resistant enteropathogenic *E. coli* strains.

For many reasons, the presence of R-factors in nonpathogenic bacteria of the gastro-intestinal tract (such as *E. coli*, *Proteus* and *Klebsiella*) can present a grave problem. First, if a carrier of such bacteria is infected by drug-sensitive *Salmonella* or *Shigella*, it is highly probable that the transfer of the R-factors will occur from the non-pathogenic resistant cells to the pathogenic sensitive ones. A second reason is the potentiality of several bacteria of the intestinal flora to produce infections of the uro-genital tract. The transmission of R-factors for multi-resistances will lead to the modification of chemotherapeutic treatments in this type of infection. Finally, it is certain that in the near future the enteric bacteria carrying R-factors will play an important role in hospital cross-infections.

The abuse of antibiotics, by providing an almost permanent selective pressure to the advantage of resistant cells, has strongly contributed to the spread of R-factors within pathogenic enterobacteria. By eliminating sensitive competitors, this selective pressure has facilitated the "infection" of several *Salmonellae* (ANDERSON, 1968). This is particularly true when multiple resistances are carried by a microorganism resistant to the drug administered. For example, if a strain carrying resistances against ampicillin, tetracycline, streptomycin and chloramphenicol is in contact with ampicillin, the progeny of this strain will retain the resistances for all the drugs.

As ANDERSON has shown, the immoderate abuse of antibiotics in animals (for prophylactic, therapeutic or feeding purposes) has ensured a very wide spread of resistant microorganisms.

Transferable Resistance in *Salmonella* Strains Isolated in Geneva between 1966 and 1970

A survey has been done on the various strains of *Salmonella* isolated in the Geneva hospitals during the last five years.

Table 12 shows the number of the strains studied, classified according to their serological group, divided into two main groups, sensitive and resistant, the latter being subdivided according to the transferability of their resistance

Table 12. Strains of *Salmonella* isolated in Geneva from 1966 to 1970

Serological group	Number of strains	Sensitive strains	Resistant strains		
			Total	transferable resistance	not transferable resistance
A	4	2	2	0	2
B	111	57	54	18	36
C	19	12	7	0	7
D	156	96	60	21	39
E	4	2	2	0	2
G	1	1			
Total	295	171 (58%)	124 (42%)	39 (31%)	85 (69%)

characters. By transferability we mean that one determinant at least is transferable to the *E. coli* strain used as recipient.

Several comments can be made on the results obtained:

a) On the whole, all the resistance patterns already described have been found, the most frequent being tetracycline resistance with 56 strains (19%), streptomycin with 24 strains (8%) and the combinations ampicillin-tetracycline and streptomycin-tetracycline.

b) Out of the 295 strains studied, 171 were sensitive, and 124 resistant to one or more drugs, giving a percentage of 42. Among the resistant strains, 31% possess transferable determinants.

c) If one considers only *S. typhi* and *S. paratyphi* strains, it may be noted that, out of 77 strains, 60 remain completely sensitive; for the 17 resistant strains no transferable determinant has been detected.

d) The most important species are *S. typhimurium* (54 strains: 28 sensitive, 26 resistant with 15 having transferable resistance), *S. dublin* (19 strains: 8 sensitive, 11 resistant with 6 having transferable resistance), *S. brandenburg* (15 strains: 4 sensitive, 11 resistant with no transferable character), *S. enteritidis* (51 strains: 38 sensitive, 13 resistant with no transferable character) and *S. panama* (27 strains: 3 sensitive, 24 resistant with 14 having transferable resistance).

These results are in accordance with those of ANDERSON (1968) and of CHABBERT et al. (1969).

If one considers the frequency of emergence of the various resistance spectra, it is evident that tetracycline resistance is largely dominant, with 45% of the resistant strains; moreover, 20% of multi-resistant strains include this antibiotic in their resistance pattern. Streptomycin resistance is less frequent, with 20% of the resistant strains. This could be explained by a decreased administration of streptomycin and an increased utilization of the different tetracyclines.

If one looks at the evolution of resistance with time, it can be seen that in 1966 we found 36% of resistant strains, 23% of which were transferable;

in 1967, 37% of resistant strains, with 39% transferable; in 1968, 36% of resistant strains with 33 % transferable; in 1969, 55 % of resistant strains with 20% transferable and in 1970, 54% of resistant strains, with 31 % transferable. Thus, although the percentage of resistant strains has increased significantly during the last two years, the percentage of those having transferable resistance remains relatively stable. But the frequency of plasmid-linked resistances is higher, since it has been demonstrated that several strains whose resistance determinants are not transferable present a mode of resistance similar to that found in strains carrying transferable R-factors.

IX. Conclusions

To conclude this paper, we shall recall the principal data supporting our hypothesis that the genetic determinants controlling bacterial resistance to antibiotics are usually extrachromosomally located.

The way in which a bacterial cell can resist any one antibiotic differs significantly according to the mode of genetic control.

With extrachromosomal determinants, the penicillins and the cephalosporins are inactivated by degradation of the molecule by penicillinases and cephalosporinases, respectively. The location of these enzymes in the cell depends on whether the synthesis is controlled by a plasmid (periplasmic enzyme, easily released in the medium) or by chromosomal genes (intracellular enzyme, bound to the cell components and not belonging to the group of enzymes located on the cell membrane). Moreover, the specific activity of the intracellular enzymes is much lower than that of the periplasmic ones. The majority of wild-type ampicillin-resistant *E. coli* (79 %) and *Salmonellae* (82 %) have a periplasmic enzyme whose synthesis is controlled by a plasmid, whether transferable or not.

For aminoglucoside antibiotics (streptomycin, kanamycin, paromomycin) resistance due to mutations on chromosomal genes results in the modification of a specific ribosomal protein, whereas plasmid-controlled resistance is due to inactivation of the drug (acetylation or phosphorylation of kanamycin, phosphorylation of paromomycin, adenylation of streptomycin and dihydrostreptomycin). All the wild-type streptomycin-resistant strains of *E. coli* and nearly 70% of the *Salmonellae* studied show a system of enzymatic inactivation for streptomycin and dihydrostreptomycin.

Resistance to chloramphenicol is due to a modification of membrane permeability in the strains for which resistance is controlled by chromosomal genes. When resistance is mediated by an extrachromosomal determinant, the antibiotic is inactivated by acetylation. This reaction is brought about by the synthesis of an acetyl-transferase, directly controlled by the plasmid. All the wild-type chloramphenicol-resistant strains of *E. coli* carry such an enzyme which is able to inactivate the drug.

As far as tetracyclines are concerned, it is not yet possible to draw positive conclusions, since we cannot differentiate between the various systems which allow the bacteria to resist this group of antibiotics. A modification of cell permeability seems to be responsible for resistance, both in the chromosomal and extrachromosomal system of control.

The origin of extrachromosomal resistance determinants still remains obscure. We do not think that they originated on the bacterial chromosome; we may suppose that a plasmidic sensitive allele is present in the wild-type strains, but its demonstration is difficult.

References

Akiba, T.: Mechanism of development of resistance in *Shigella*. Proc. 15th Gen. Meeting of the Jap. Med. Assoc. 5, 299–305 (1959); quoted by Watanabe (1963).

Akiba, T., Koyama, K., Ishiki, Y., Kimura, S., Fukushima, T.: On the mechanism of the development of multiple drug-resistant clones of *Shigella*. Jap. J. Microbiol. **4**, 219–227 (1960).

Akiba, T., Yokota, T.: Studies on the mechanism of transfer of drug-resistance in bacteria. 22. Influences of chloramphenicol and tetracycline on the ^{14}C-amino acid incorporation by ribosomes isolated from the drug-sensitive strain and the multiple resistant strain of *E. coli*. Med. Biol. (Tokyo) **64**, 34 (1962); quoted by Okamoto and Mizuno (1964).

Anderson, E. S.: A rapid screening test for transfer factors in drug-sensitive *Enterobacteriaceae*. Nature (Lond.) **208**, 1016–1017 (1965a).

Anderson, E. S.: Origin of transferable drug-resistance factors in the *Enterobacteriaceae*. Brit. med. J. **1965b II**, 1289–1291.

Anderson, E. S.: Influence of the *Δ* transfer factor on the phage sensitivity of *Salmonellae*. Nature (Lond.) **212**, 795–799 (1966).

Anderson, E. S.: Facteurs de transfert et résistances aux antibiotiques chez les Entérobactéries. Ann. Inst. Pasteur **112**, 547–563 (1967).

Anderson, E. S.: The ecology of transferable drug resistance in the *Enterobacteria*. Ann. Rev. Microbiol. **22**, 131–180 (1968).

Anderson, E. S., Datta, N.: Resistance to penicillins and its transfer in *Enterobacteriaceae*. Lancet **1965 I**, 407–409.

Anderson, E. S., Kelemen, M. V., Jones, C. M., Pitton, J. S.: Study of the association of resistance to two drugs in a transferable determinant in *Salmonella typhimurium*. Genet. Res. **11**, 119–124 (1968).

Anderson, E. S., Lewis, M. J.: Drug resistance and its transfer in *Salmonella typhimurium*. Nature (Lond.) **206**, 579–583 (1965a).

Anderson, E. S., Lewis, M. J.: Characterization of a transfer factor associated with drug resistance in *Salmonella typhimurium*. Nature (Lond.) **208**, 843–849 (1965b).

Anderson, T. F., Wollman, E., Jacob, F.: Sur les processus de conjugaison et de recombinaison chez *E. coli*. III. Aspects morphologiques en microscopie électronique. Ann. Inst. Pasteur **93**, 450–455 (1957).

Anderson, W. F., Gorini, L., Breckenridge, L.: Role of ribosomes in streptomycin-activated suppression. Proc. nat. Acad. Sci. (Wash.) **54**, 1076–1083 (1965).

Apirion, D., Schlessinger, D.: The loss of phenotypic suppression in streptomycin-resistant mutants. Proc. nat. Acad. Sci. (Wash.) **58**, 206–212 (1967a).

Apirion, D., Schlessinger, D.: Reversion from streptomycin dependence in *Escherichia coli* by a further change in the ribosome. J. Bact. **94**, 1275–1276 (1967b).

Apirion, D., Schlessinger, D.: Coresistance to neomycin and kanamycin by mutations in an *Escherichia coli* locus that affects ribosomes. J. Bact. **96**, 768–776 (1968).

Apirion, D., Schlessinger, D., Phillips, S., Sypherd, P.: *Escherichia coli:* reversion from streptomycin dependence, a mutation in a specific 30S ribosomal protein. J. molec. Biol. **43**, 327–329 (1969).

ASHESHOV, E. H.: Chromosomal location of the genetic elements controlling penicillinase production in a strain of *Staphylococcus aureus*. Nature (Lond.) **210**, 804–806 (1966a).

ASHESHOV, E. H.: Loss of antibiotic resistance in *Staphylococcus aureus* resulting from growth at high temperature. J. gen. Microbiol. **42**, 403–410 (1966b).

BARBER, M.: Drug resistance of *Staphylococci* with special reference to penicillinase production. In: Ciba Foundation Symposium on Drug Resistance in Micro-Organisms, p. 262–274. Ed. by G. E. WOLSTENHOLME and C. M. O'CONNOR. London: J. & A. Churchill 1957.

BARBER, M.: Coagulase-positive *Staphylococci* resistant to benzylpenicillin, methicillin and other penicillins. In: Ciba Foundation Study Group No 13: Resistance of bacteria to the penicillins, p. 89–102. London: J. & A. Churchill 1962.

BATCHELOR, F. R., CHAIN, E. B., RICHARDS, M., ROLINSON, G. N.: 6-aminopenicillanic acid. VI. Formation of 6-aminopenicillanic acid from penicillins by enzymic hydrolysis. Proc. roy. Soc. B **154**, 522–531 (1961).

BENBOUGH, J., MORRISON, G. A.: Bacteriostatic actions of some tetracyclines. J. Pharm. Pharmacol. **17**, 409–422 (1965).

BOMAN, H. G., ERIKSSON-GRENNBERG, K. G., NORMARK, S., MATSSON, E.: Resistance of *Escherichia coli* to penicillins. IV. Genetic study of mutants resistant to D,L-ampicillin concentrations of 100 µg/ml. Genet. Res. **12**, 169–185 (1968).

BOUANCHAUD, D.: Recherche colorimétrique de l'estérification du chloramphénicol par les entérobactéries et les staphylocoques porteurs de caractères de résistance transférables. Ann. Inst. Pasteur **113**, 59–66 (1967).

BOUANCHAUD, D., SCAVIZZI, M. R., CHABBERT, Y. A.: Elimination by ethidium bromide of antibiotic resistance in *Enterobacteria* and *Staphylococci*. J. gen. Microbiol. **54**, 417–425 (1968).

BOWNE, S. W., ROGERS, P.: Accumulation of repressor for ornithine transcarbamylase synthesis in *Escherichia coli* mediated by chloramphenicol. Biochim. biophys. Acta (Amst.) **76**, 600–613 (1963).

BRINTON, C. C., JR.: The structure, function, synthesis and genetic control of bacterial pili and a molecular model for DNA and RNA transport in gram-negative bacteria. Trans. N.Y. Acad. Sci. **27**, 1003–1054 (1965).

BRINTON, C. C., JR., GEMSKI, P., CARNAHAN, J.: A new type of bacterial pilus genetically controlled by the fertility factor of *E. coli* K12 and its role in chromosome transfer. Proc. nat. Acad. Sci. (Wash.) **52**, 776–783 (1964).

BROCK, T. D.: Streptomycin. In: Biochemical studies of antimicrobial drugs. Symp. Soc. gen. Microbiol. **16**, 131–168 (1966).

BRODY, T. M., HURWITZ, R., BAIN, J. A.: Magnesium and the effects of the tetracycline antibiotics on oxidative processes in mitochondria. Antibiot. et Chemother. (Basel) **4**, 864–870 (1954).

CANNON, M., KRUG, R., GILBERT, W.: The binding of s-RNA by *Escherichia coli* ribosomes. J. molec. Biol. **7**, 360–378 (1963).

CAVALLI, L. L., MACCACARO, G. A.: Polygenic inheritance of drug-resistance in the bacterium *Escherichia coli*. Heredity **6**, 311–331 (1952).

CHABBERT, Y. A., BAUDENS, J. G.: Transmissible resistance to six groups of antibiotics in *Salmonella* infections. Antimicrobial Agents and Chemother., p. 380–383 (1965).

CHABBERT, Y. A., BAUDENS, J. G., BOUANCHAUD, D. H.: Medical aspects of transferable drug resistance. In: Ciba Foundation Symposium on Drug Resistance in Micro-Organisms, p. 227–239. Ed. by G. E. WOLSTENHOLME and C. M. O'CONNOR. London: J. & A. Churchill 1969.

CHABBERT, Y. A., BAUDENS, J. G., GERBAUD, G. R.: Variations sous l'influence de l'acriflavine et transduction de la résistance à la kanamycine et au chloramphénicol chez les Staphylocoques. Ann. Inst. Pasteur **107**, 678–690 (1964).

CHUIT, C. F.: Résistance à la tétracycline et au chloramphénicol chez *Escherichia coli* K12. Thèse, Genève 1968.

CHUIT, C. F., PITTON, J. S.: Non-transferable tetracycline resistance in *Escherichia coli* K12. Chemotherapy **14**, 253–257 (1969).

CLOWES, R. C., HAYES, W.: Experiments in microbial genetics. Oxford-Edinburgh: Blackwell 1968.

COLE, M., SUTHERLAND, R.: The role of penicillin acylase in the resistance of gram-negative bacteria to penicillins. J. gen. Microbiol. **42**, 345–356 (1966).

Connamacher, R. H., Mandel, H. G.: Binding of tetracycline to the 30S ribosomes and to polyuridilic acid. Biochim. biophys. Acta (Amst.) **166**, 475–486 (1968).
Couturier, M., Desmet, L., Thomas, R.: High pleiotrophy of streptomycin mutations in *Escherichia coli*. Biochem. biophys. Res. Commun. **16**, 244–248 (1964).
Cox, E. C., White, J. R., Flaks, J. G.: Streptomycin action on the ribosome. Proc. nat. Acad. Sci. (Wash.) **51**, 703–709 (1964).
Cundliffe, E., McQuillen, K.: Bacterial protein synthesis: the effects of antibiotics. J. molec. Biol. **30**, 137–146 (1967).
Das, H. K., Goldstein, A., Kanner, L. C.: Inhibition by chloramphenicol of the growth of nascent protein chains in *Escherichia coli*. Molec. Pharmacol. **2**, 158–170 (1966).
Datta, N.: Transmissible drug resistance in an epidemic strain of *Salmonella typhimurium*. J. Hyg. (Lond.) **60**, 301–310 (1962).
Datta, N., Kontomichalou, P.: Penicillinase synthesis controlled by infectious R factors in *Enterobacteriaceae*. Nature (Lond.) **208**, 239–241 (1965).
Datta, N., Richmond, M. H.: The purification and properties of a penicillinase whose synthesis is mediated by an R-factor in *Escherichia coli*. Biochem. J. **98**, 204–209 (1966).
Davies, J.: Streptomycin and the genetic code. Cold Spr. Harb. Symp. quant. Biol. **31**, 665–670 (1966).
Davies, J., Davis, B. D.: Misreading of RNA codewords induced by aminoglucoside antibiotics: the effect of drug concentration. J. biol. Chem. **243**, 3312–3316 (1968).
Day, L. E.: Tetracycline inhibition of cell-free protein synthesis. I. Binding of tetracycline to components of the system. J. Bact. **91**, 1917–1923 (1966a).
Day, L. E.: Tetracycline inhibition of cell-free protein synthesis. II. Effect of the binding of tetracycline to the components of the system. J. Bact. **92**, 197–203 (1966b).
Demerec, M.: Origin of bacterial resistance to antibiotics. J. Bact. **56**, 63–74 (1948).
Dettori, R., Maccacaro, G. A., Piccinin, G. L.: Sex-specific bacteriophages of *Escherichia coli* K12. G. Microbiol. **9**, 141–150 (1961).
Dubin, D. T.: Some effects of streptomycin on RNA metabolism in *Escherichia coli*. J. molec. Biol. **8**, 749–767 (1964).
Dunsmoor, C. L., Pim, K. L., Sherris, J. C.: Observations on the inactivation of chloramphenicol by chloramphenicol-resistant *Staphylococci*. Antimicrobial Agents and Chemother., p. 500–506 (1963).
Elliott, W. H.: The effects of antimicrobial agents on deoxyribonucleic acid polymers. Biochem. J. **86**, 562–567 (1963).
Eriksson-Grennberg, K. G.: Resistance of *Escherichia coli* to penicillins. II. An improved mapping of the *ampA* gene. Genet. Res. **12**, 147–156 (1968).
Eriksson-Grennberg, K. G., Boman, H. G., Torbjörn-Jansson, J. A., Thoren, S.: Resistance of *Escherichia coli* to penicillins. I. Genetic study of some ampicillin-resistant mutants. J. Bact. **90**, 54–62 (1965).
Falkow, S., Citarella, R. V., Wohlhieter, J. A., Watanabe, T.: The molecular nature of R factors. J. molec. Biol. **17**, 102–116 (1966).
Flaks, J. G., Cox, E. C., Witting, M. L., White, J. R.: Polypeptide synthesis with ribosomes from streptomycin-resistant and dependent *E. coli*. Biochem. biophys. Res. Commun. **7**, 390–393 (1962).
Fleming, P. C., Goldner, M., Glass, D. G.: Observations on the nature, distribution and significance of cephalosporinase. Lancet **1963 I**, 1399–1401.
Franklin, T. J.: The inhibition of incorporation of leucine into protein of cell-free systems from rat liver and *Escherichia coli* by chlortetracycline. Biochem. J. **87**, 449–453 (1963).
Franklin, T. J.: Resistance of *Escherichia coli* to tetracyclines. Changes in permeability to tetracyclines in *Escherichia coli* bearing transferable resistance factors. Biochem. J. **105**, 371–378 (1967).
Franklin, T. J., Godfrey, A.: Resistance of *Escherichia coli* to tetracycline. Biochem. J. **94**, 54–60 (1965).
Gale, E. F., Folkes, J. P.: The assimilation of amino acids by bacteria. 15. Action of antibiotics on nucleic acid and protein synthesis in *Staphylococcus aureus*. Biochem. J. **53**, 493–498 (1953).
Geronimus, L. H., Cohen, S.: Induction of staphylococcal penicillinase. J. Bact. **73**, 28–34 (1957).

GOTTESMAN, M. E.: Reaction of ribosome-bound peptidyl transfer ribonucleic acid with aminoacyl transfer ribonucleic acid or puromycin. J. biol. Chem. **242**, 5564–5571 (1967).

GROS, F., DUBERT, J. M., TISSIERES, A., BOURGEOIS, S., MICHELSON, M., SOFFER, R., LEGAULT, L.: Regulation of metabolic breakdown and synthesis of messenger RNA in bacteria. Cold Spr. Harb. Symp. quant. Biol. **28**, 299–313 (1963).

HAHN, F. E.: Chloramphenicol. In: Antibiotics, vol. I, p. 308–330. Ed. by D. GOTTLIEB and P. D. SHAW. Berlin-Heidelberg-New York: Springer 1967.

HAHN, F. E., WOLFE, A. D.: Mode of action of chloramphenicol. VIII. Resemblance between labile chloramphenicol-ribonucleic acid and deoxyribonucleic acid of *Bacillus cereus*. Biochem. biophys. Res. Commun. **6**, 464–468 (1961).

HAMILTON-MILLER, J. M. T.: Penicillinase from *Klebsiella aerogenes*. Biochem. J. **87**, 209–214 (1963a).

HAMILTON-MILLER, J. M. T.: Inducible penicillinase in *Proteus morgani*. Biochem. biophys. Res. Commun. **13**, 43–48 (1963b).

HARADA, K., KAMEDA, M., SUZUKI, M., EGAWA, R., MITSUHASHI, S.: Studies on the drug resistance of enteric bacteria. 10. Relation between transmissible drug resistance (R) factor and fertility (F) factor in *E. coli* strain K12. Jap. J. exp. Med. **31**, 291–299 (1961).

HARMON, S. A., BALDWIN, J. N.: Nature of the determinant controlling penicillinase production in *Staphylococcus aureus*. J. Bact. **87**, 593–597 (1964).

HARWOOD, J. H., SMITH, D. H.: Resistance factor-mediated streptomycin resistance. J. Bact. **97**, 1262–1271 (1969).

HASHIMOTO, H., HIROTA, Y.: Gene recombination and segregation of resistance factor R in *Escherichia coli*. J. Bact. **91**, 51–62 (1966).

HASHIMOTO, H., KONDO, K., MITSUHASHI, S.: Elimination of penicillin resistance of *Staphylococcus aureus* by treatment with acriflavine. J. Bact. **88**, 261–262 (1964).

HENNESSEY, T. D.: Inducible β-lactamase in *Enterobacter*. J. gen. Microbiol. **49**, 277–285 (1967).

HIEROWSKI, M.: Inhibition of protein synthesis by chlortetracycline in the *E. coli in vitro* system. Proc. nat. Acad. Sci. (Wash.) **53**, 594–599 (1965).

HOLMES, I. A., WILD, D. G.: Consequences of inhibition of *Escherichia coli* by tetracycline antibiotics. Nature (Lond.) **210**, 1047–1048 (1966).

IZAKI, K., ARIMA, K.: Disappearance of oxytetracycline accumulation in the cells of multiple drug-resistant *Escherichia coli*. Nature (Lond.) **200**, 384–385 (1963).

IZAKI, K., ARIMA, K.: Effect of various conditions on accumulation of oxytetracycline in *Escherichia coli*. J. Bact. **89**, 1335–1339 (1965).

IZAKI, K., KIUCHI, K., ARIMA, K.: Specificity and mechanism of tetracycline resistance in a multiple drug-resistant strain of *Escherichia coli*. J. Bact. **91**, 628–633 (1966).

JACK, G. W., RICHMOND, M. H.: A comparative study of eight distinct β-lactamases synthesized by gram-negative bacteria. J. gen. Microbiol. **61**, 43–61 (1970).

JACOB, F., MONOD, J.: Genetic regulatory mechanisms in the synthesis of proteins. J. molec. Biol. **3**, 318–356 (1961).

JACOB, F., SCHAEFFER, P., WOLLMAN, E. L.: Episomic elements in bacteria. Symp. Soc. gen. Microbiol. **10**, 67–91 (1960).

JACOB, F., ULLMAN, A., MONOD, J.: Délétions fusionnant l'opéron lactose et un opéron purine chez *E. coli*. J. molec. Biol. **13**, 704–719 (1965).

JACOB, F., WOLLMAN, E. L.: Sur les processus de conjugaison et de recombinaison génétique chez *E. coli*. I. L'induction par conjugaison ou induction zygotique. Ann. Inst. Pasteur **91**, 486–510 (1956).

JAGO, M., MIGLIACI, A., ABRAHAM, E. P.: Production of cephalosporinase by *Pseudomonas pyocyanea*. Nature (Lond.) **199**, 375 (1963).

JAROLMEN, H., BONDI, A., CROWELL, R. L.: Transduction of *Staphylococcus aureus* to tetracycline resistance *in vivo*. J. Bact. **89**, 1286–1290 (1965).

JONES, J. G., MORRISON, G. A.: The bacteriostatic actions of tetracycline and oxytetracycline. J. Pharm. (Lond.) **14**, 808–824 (1962).

JULIAN, G. R.: ^{14}C-lysine peptides synthetized in an *in vitro Escherichia coli* system in the presence of chloramphenicol. J. molec. Biol. **12**, 9–16 (1965).

KABINS, S. A., COHEN, S.: Resistance transfer factor in *Enterobacteriaceae*. New Engl. J. Med. **275**, 248–252 (1966).

KASATIYA, S. S., BALDWIN, J. N.: Nature of the determinant of tetracycline resistance in *Staphylococcus aureus*. Canad. J. Microbiol. **13**, 1079–1086 (1967).

KITAMOTO, O., KASAI, N., FUKAYA, K., KAWASHIMA, A.: Drug-sensitivity of the *Shigella* strains isolated in 1955. J. Jap. Ass. Infect. Dis. **30**, 403–404 (1956); quoted by WATANABE (1963).

KNOX, R.: Different types of resistance to different penicillins. In: Ciba Foundation Study Group No 13: Resistance of bacteria to the penicillins, p. 76–83. London: J. & A. Churchill 1962.

KNOX, R., SMITH, J. T.: Antibacterial activity, penicillinase stability and inducing ability of different penicillins. J. gen. Microbiol. **29**, 471–479 (1962).

KONDO, S., OKANISHI, M., UTAHARA, R., UMEZAWA, H.: Inactivation of aminoglycosidic antibiotics by resistant organisms. Jap. J. med. Sci. Biol. **21**, 221–223 (1968).

KRCMERY, V., KELLEN, J.: Changes in some enzymes of bacterial electron transport accompanying development of resistance to oxytetracycline. J. Bact. **92**, 1264–1266 (1966).

KROON, A. M.: Protein synthesis in mitochondria. III. On the effects of inhibitors on the incorporation of amino acids into protein by intact mitochondria and digitonin fractions. Biochim. biophys. Acta (Amst.) **108**, 275–284 (1965).

KUCAN, Z., F. LIPMANN: Differences in chloramphenicol sensitivity of cell-free amino acid polymerization systems. J. biol. Chem. **239**, 516–520 (1964).

KURLAND, C. G., NOMURA, M., WATSON, J. D.: The physical properties of the chloromycetin particles. J. molec. Biol. **4**, 388–394 (1962).

KUSCHNER, D. J.: The basis of chloramphenicol resistance in *Pseudomonas fluorescens*. Arch. Biochem. **58**, 347–355 (1955).

KUWANO, M., ISHIKAWA, M., ENDO, H.: Su-II-specific restriction of amber suppression by mutation to streptomycin resistance. J. molec. Biol. **33**, 513–516 (1968).

LASKIN, A. I., CHAN, W. M.: Inhibition by tetracyclines of polyuridilic acid-directed phenylalanine incorporation in *Escherichia coli* cell-free systems. Biochem. biophys. Res. Commun. **14**, 137–142 (1964).

LAST, J. A., IZAKI, K., SNELL, J. F.: The failure of tetracycline to bind to *Escherichia coli* ribosomes. Biochim. biophys. Acta (Amst.) **103**, 532–534 (1965).

LAST, J. A., IZAKI, K., SNELL, J. F.: The resistance of *Escherichia coli* to oxytetracycline. Canad. J. Microbiol. **15**, 1077–1083 (1969).

LEBEK, G.: Über die Entstehung mehrfachresistenter Salmonellen. Ein experimenteller Beitrag. Zbl. Bakt., I. Abt. Orig. **188**, 494–505 (1963a).

LEBEK, G.: Übertragung der Mehrfachresistenz gegen Antibiotika und Chemotherapeutika von *E. coli* auf andere Species gramnegativer Bakterien. Experimenteller Beitrag. Zbl. Bakt., I. Abt. Orig. **189**, 213–223 (1963b).

LEBEK, G.: Die Übertragung der Mehrfachresistenz gegen Antibiotika und Chemotherapeutika in ihrer Bedeutung für den Hospitalismus mit mehrfachresistenten gramnegativen Darmbakterien. Zbl. Bakt., I. Abt. Orig. **191**, 387–395 (1963c).

LEDERBERG, E. M., CAVALLI-SFORZA, L., LEDERBERG, J.: Interaction of streptomycin and a suppressor for galactose fermentation in *E. coli* K12. Proc. nat. Acad. Sci. (Wash.) **51**, 678–682 (1964).

LEDERBERG, J.: Streptomycin resistance: a genetically recessive mutation. J. Bact. **61**, 549–550 (1951).

LEDERBERG, J.: Bacterial protoplasts induced by penicillin. Proc. nat. Acad. Sci. (Wash.) **42**, 574–577 (1956).

LEDERBERG, J., LEDERBERG, E. M.: Replica plating and indirect selection of bacterial mutants. J. Bact. **63**, 399–406 (1952).

LEDERBERG, S.: Suppression of multiplication of heterologous bacteriophages in lysogenic bacteria. Virology **3**, 496–513 (1957).

LEWIS, M. J.: Multiple transmissible drug resistance in an outbreak of *Shigella flexneri* infection. Lancet **1967 II**, 953–956.

LIKOVER, T. E., KURLAND, C. G.: Ribosomes from a streptomycin-dependent strain of *Escherichia coli*. J. molec. Biol. **25**, 497–504 (1967).

LINDQVIST, R. CHR., NORDSTRÖM, K.: Resistance of *Escherichia coli* to penicillins. VII. Purification and characterization of a penicillinase mediated by the R factor R1. J. Bact. **101**, 232–239 (1970).

LINDSTRÖM, E. B., BOMAN, H. G., STEELE, B. B.: Resistance of *Escherichia coli* to penicillins. VI. Purification and characterization of the chromosomally mediated penicillinase present in *ampA*-containing strains. J. Bact. **101**, 218–231 (1970).
LOEB, T.: Isolation of a bacteriophage specific for the F^+ and Hfr mating types of *E. coli*. Science **131**, 932–933 (1960).
LOOMIS, W. F.: On the mechanism of action of aureomycin. Science **111**, 474 (1950).
LUCAS-LENARD, J., HAENNI, A.: Requirement of guanosine-5'-triphosphate for ribosomal binding of aminoacyl-s-RNA. Proc. nat. Acad. Sci. (Wash.) **59**, 554–560 (1968).
LURIA, S. E., DELBRÜCK, M.: Mutations of bacteria from virus sensitivity to virus resistance. Genetics **28**, 491–511 (1943).
LUZZATTO, L., SCHLESSINGER, D., APIRION, D.: *Escherichia coli:* high resistance or dependence on streptomycin produced by the same allele. Science **161**, 478–479 (1968).
MACUCH, P., KRCMERY, V., PARRAKOVA, E., SECKAROVA, A.: Transmissibility of tetracycline resistance from antibiotic-selected mutants *E. coli* to *S. typhimurium* and vice-versa. In: Vth Internat. Congr. of Chemotherapy, Vienna 1967, vol. IV, p. 243–246. Vienna: Verlag der Wiener Med. Akademie 1967.
MANDELSTAM, J., ROGERS, H. J.: The incorporation of amino acids into the cell wall mucopeptide of *Staphylococci* and the effects of antibiotics on the process. Biochem. J. **72**, 654–662 (1959).
MARKOVITZ, A., BAKER, B.: Suppression of radiation sensitivity and capsular polysaccharide synthesis in *Escherichia coli* K12 by ochre suppressors. J. Bact. **94**, 388–395 (1967).
MAXWELL, I. H.: Studies of the binding of tetracycline to ribosomes *in vitro*. Molec. Pharmacol. **4**, 25–37 (1968).
MAY, J. W., HOUGHTON, R. H., PERRET, C. J.: The effect of growth at elevated temperatures on some heritable properties of *Staphylococcus aureus*. J. gen. Microbiol. **37**, 157–169 (1964).
MERKEL, J. R., STEERS, E.: Relationship between "chloramphenicol reductase activity" and chloramphenicol resistance in *Escherichia coli*. J. Bact. **66**, 389–396 (1953).
MEYNELL, E., DATTA, N.: The relation of resistance transfer factors to the F-factor (sex-factor) of *Escherichia coli* K12. Genet. Res. **7**, 134–140 (1966a).
MEYNELL, E., DATTA, N.: The nature and incidence of conjugation factors in *Escherichia coli*. Genet. Res. **7**, 141–148 (1966b).
MEYNELL, E., DATTA, N.: Mutant drug-resistant factors of high transmissibility. Nature (Lond.) **214**, 885–887 (1967).
MISE, K., SUZUKI, Y.: Temperature-sensitive chloramphenicol acetyltransferase from *Escherichia coli* carrying mutant R factors. J. Bact. **95**, 2124–2130 (1968).
MITSUHASHI, S.: Epidemiological and genetic study of drug resistance in *Staphylococcus aureus*. Jap. J. Microbiol. **2**, 49–68 (1966).
MITSUHASHI, S., HARADA, K., HASHIMOTO, H.: Multiple resistance of enteric bacteria and transmission of drug-resistance to other strains by mixed cultivation. Jap. J. exp. Med. **30**, 179–184 (1960).
MITSUHASHI, S., HARADA, K., KAMEDA, M.: On the drug-resistance of enteric bacteria. 6. Spontaneous and artificial elimination of transmissible drug-resistance factors. Jap. J. exp. Med. **31**, 119–123 (1961a).
MITSUHASHI, S., HARADA, K., KAMEDA, M.: Elimination of transmissible drug-resistance by treatment with acriflavine. Nature (Lond.) **189**, 947 (1961b).
MIYAMURA, S.: Inactivation of chloramphenicol by chloramphenicol resistant bacteria. J. pharm. Sci. **53**, 604–607 (1964).
MONRO, R. E.: Catalysis of peptide bond formation by 50S ribosomal subunits from *Escherichia coli*. J. molec. Biol. **26**, 147–151 (1967).
MONRO, R. E., MARCKER, K. A.: Ribosome-catalyzed reaction of puromycin with a formylmethionine-containing oligonucleotide. J. molec. Biol. **25**, 347–350 (1967).
MOORE, P. B., TRAUT, R. R., NOLLER, H., PEARSON, P., DELIUS, H.: Ribosomal proteins of *Escherichia coli*. II. Proteins from the 30S subunit. J. molec. Biol. **31**, 441–461 (1968).
MOYED, H. S.: Induced phenotypic resistance to an antimetabolite. Science **131**, 1449 (1960).
NAKAMOTO, T., CONWAY, T. W., ALLENDE, J. E., SPYRIDES, C. P., LIPMANN, F.: Formation of peptide bonds. I. Peptide formation from aminoacyl-s-RNA. Cold Spr. Harb. Symp. quant. Biol. **28**, 227–231 (1963).

NEU, H. C.: Effect of β-lactamase location in *Escherichia coli* on penicillin synergy. Appl. Microbiol. **17**, 783–786 (1969).

NEU, H. C., CHOU, J.: Release of surface enzymes in *Enterobacteriaceae* by osmotic shock. J. Bact. **94**, 1934–1945 (1967).

NOMURA, M., HOSOKAWA, K.: Biosynthesis of ribosomes: fate of chloramphenicol particles and of pulse-labelled RNA in *Escherichia coli*. J. molec. Biol. **12**, 242–265 (1965).

NORDSTRÖM, K., ERIKSSON-GRENNBERG, K. G., BOMAN, H. G.: Resistance of *Escherichia coli* to penicillins. III. *AmpB*, a locus affecting episomally and chromosomally mediated resistance to ampicillin and chloramphenicol. Genet. Res. **12**, 157–168 (1968).

NOSSAL, N. G., HEPPEL, L. A.: The release of enzymes by osmotic shock from *Escherichia coli* in exponential phase. J. biol. Chem. **241**, 3055–3062 (1966).

NOVICK, R. P.: Micro-iodometric assay for penicillinase. Biochem. J. **83**, 236–240 (1962).

NOVICK, R. P.: Analysis by transduction of mutations affecting penicillinase formation in *Staphylococcus aureus*. J. gen. Microbiol. **33**, 121–136 (1963).

NOVICK, R. P.: Extrachromosomal inheritance in bacteria. Bact. Rev. **33**, 210–263 (1969).

NOVICK, R. P., RICHMOND, M. H.: Nature and interactions of the genetic elements governing penicillinase synthesis in *Staphylococcus aureus*. J. Bact. **90**, 467–480 (1965).

NOVICK, R. P., MORSE, S. I.: *In vivo* transmission of drug resistance factors between strains of *Staphylococcus aureus*. J. exp. Med. **125**, 45–59 (1967).

OCHIAI, K., YAMANAKA, T., KIMURA, K., SAWADA, O.: Studies of inheritance of drug resistance between *Shigella* strains and *Escherichia coli* strains. Nippon Iji Shimpo **1861**, 34–46 (1959); quoted by WATANABE (1963).

OKAMOTO, S., MIZUNO, D.: Inhibition by chloramphenicol of protein synthesis in the cell-free system of a chloramphenicol-resistant strain of *Escherichia coli*. Nature (Lond.) **195**, 1022–1023 (1962).

OKAMOTO, S., MIZUNO, D.: Mechanism of chloramphenicol and tetracycline resistance in *Escherichia coli*. J. gen. Microbiol. **35**, 125–133 (1964).

OKAMOTO, S., SUZUKI, Y.: Chloramphenicol-, dihydrostreptomycin-, and kanamycin-inactivating enzymes from multiple drug-resistant *Escherichia coli* carrying episome "R". Nature (Lond.) **208**, 1301–1303 (1965).

OKAMOTO, S., SUZUKI, Y., MISE, K., NAKAYA, R.: Occurrence of chloramphenicol-acetylating enzymes in various gram-negative bacilli. J. Bact. **94**, 1616–1622 (1967).

OZAKI, M., MIZUSHIMA, S., NOMURA, M.: Identification and functional characterization of the protein controlled by the streptomycin-resistant locus in *E. coli*. Nature (Lond.) **222**, 333–339 (1969).

PARK, J. T.: Uridine-5′-pyrophosphate derivatives. I. Isolation from *Staphylococcus aureus*. J. biol. Chem. **194**, 877–884 (1952).

PERCIVAL, A., BRUMFITT, W., LOUVOIS, J. DE: The role of penicillinase in determining natural and acquired resistance of gramnegative bacteria to penicillins. J. gen. Microbiol. **32**, 77–89 (1963).

PERRET, C. J.: Iodometric assay of penicillinase. Nature (Lond.) **174**, 1012–1013 (1954).

PIFFARETTI, J. C., ALLET, B., PITTON, J. S.: Analogy between *in vivo* and *in vitro* biological effect of chloramphenicol and its acetylated derivatives. FEBS Letters **11**, 26–28 (1970).

PIFFARETTI, J. C., PITTON, J. S.: Chloramphenicol acetylation in whole cells of *Escherichia coli* carrying R-factors. Characterization and kinetic studies. Chemotherapy **15**, 84–98 (1970).

PIGUET, J. D., PITTON, J. S.: Standardisation de l'interprétation des dosages microbiologiques d'antibiotiques. Emploi d'une calculatrice électronique. I. Présentation du problème. Pharm. Acta Helv. **43**, 713–725 (1968).

PITTON, J. S., ANDERSON, E. S.: The inhibitory action of transfer factors on lysis of *Escherichia coli* K12 by phages μ2 and 2. Genet. Res. **16**, 215–224 (1970).

POLLOCK, M. R.: Penicillinase. In: Ciba Foundation Study Group No 13: Resistance of bacteria to the penicillins, p. 56–70. Ed. by G. E. WOLSTENHOLME and C. M. O'CONNOR. London: J. & A. Churchill 1962.

POSTON, S. M.: Cellular location of the genes controlling penicillinase production and resistance to streptomycin and tetracycline in a strain of *Staphylococcus aureus*. Nature (Lond.) **210**, 802–804 (1966).

RAMSEY, H. H.: Protein synthesis as a basis for chloramphenicol-resistance in *Staphylococcus aureus*. Nature (Lond.) **182**, 602–603 (1958).

RASSEKH, M., PITTON, J. S.: Streptomycin resistance in some wild-type strains of *Enterobacteriaceae*. Chemotherapy (in press).
REEVE, E. C. R.: Genetic analysis of some mutations causing resistance to tetracycline in *Escherichia coli* K12. Genet. Res. **11**, 303–309 (1968).
REEVE, E. C. R., DOHERTY, P.: Linkage relationship of two genes causing partial resistance to chloramphenicol in *Escherichia coli*. J. Bact. **96**, 1450–1451 (1968).
REEVE, E. C. R., SUTTIE, D. R.: Chromosomal location of a mutation causing chloramphenicol resistance in *Escherichia coli* K12. Genet. Res. **11**, 97–104 (1968).
RENDI, R., OCHOA, S.: Effect of chloramphenicol on protein synthesis in cell-free preparations of *Escherichia coli*. J. biol. Chem. **237**, 3711–3713 (1962).
RITZ, H. L., BALDWIN, J. N.: Transduction of capacity to produce staphylococcal penicillinase. Proc. Soc. exp. Biol. (N.Y.) **107**, 678–680 (1961).
ROGERS, H. J.: Mode of action of the penicillins. In: Ciba Foundation Study Group No 13: Resistance of bacteria to the penicillins, p. 25–43. London: J. & A. Churchill 1962.
ROLINSON, G. N., STEVENS, S.: Microbiological studies on a new broad-spectrum penicillin, "Penbritin". Brit. med. J. **1961 II**, 191–196.
ROSENKRANZ, H. S.: Basis of streptomycin resistance in *Escherichia coli* with a "multiple drug resistance" episome. Biochim. biophys. Acta (Amst.) **80**, 342–345 (1964).
ROWND, R., NAKAYA, R., NAKAMURA, A.: Molecular nature of the drug-resistance factors of the *Enterobacteriaceae*. J. molec. Biol. **17**, 376–393 (1966).
SABATH, L. D., GERSTEIN, D. A., LODER, P. B., FINLAND, M. F.: Independent segregation of chloramphenicol resistance in *Staphylococcus aureus*. Antimicrobial Agents and Chemother., p. 264–270 (1967).
SABATH, L. D., JAGO, M., ABRAHAM, E. P.: Cephalosporinase and penicillinase activity of a β-lactamase from *Pseudomonas pyocyanea*. Biochem. J. **96**, 739–752 (1965).
SARKAR, S., THACH, R. E.: Inhibition of formylmethionyl-transfer RNA binding to ribosomes by tetracycline. Proc. nat. Acad. Sci. (Wash.) **60**, 1479–1486 (1968).
SAWAI, T., MITSUHASHI, S., YAMAGISHI, S.: Drug resistance of enteric bacteria. XIV. Comparison of β-lactamases in gram-negative rod bacteria resistant to α-aminobenzylpenicillin. Jap. J. Microbiol. **12**, 423–434 (1968a).
SAWAI, T., MITSUHASHI, S., YAMAGISHI, S.: Comparison of the chromosomal and extrachromosomal genetic determinants controlling staphylococcal penicillinase production. Jap. J. Microbiol. **12**, 531–533 (1968b).
SAZ, A. K., MARTINEZ, L. M.: Enzymatic basis of resistance to aureomycin. II. Inhibition of electron transport in *Escherichia coli* by aureomycin. J. biol. Chem. **233**, 1020–1022 (1958).
SAZ, A. K., SLIE, R. B.: Inhibition of organic nitro-reductase by aureomycin in cell-free extracts. II. Co-factor requirements for the nitroreductase enzyme complex. Arch. Biochem. **51**, 5–16 (1954).
SHAW, W. V.: Enzymatic chloramphenicol acetylation and R factor induced antibiotic resistance in *Enterobacteriaceae*. Antimicrobial Agents and Chemother., p. 221–226 (1966).
SHAW, W. V.: The enzymatic acetylation of chloramphenicol by extracts of R factor-resistant *Escherichia coli*. J. biol. Chem. **242**, 687–693 (1967).
SHAW, W. V., BRODSKY, R. F.: Chloramphenicol resistance by enzymatic acetylation: comparative aspects. Antimicrobial Agents and Chemother., p. 257–263 (1967).
SHAW, W. V., BRODSKY, R. F.: Characterization of chloramphenicol acetyltransferase from chloramphenicol-resistant *Staphylococcus aureus*. J. Bact. **95**, 28–36 (1968).
SMITH, D. H.: R-factor mediated resistance to new aminoglycoside antibiotics. Lancet **1967 I**, 252–254.
SMITH, H. W., HALLS, S.: Observations on infective drug resistance in Britain. Brit. med. J. **1966 I**, 266–269.
SMITH, J. T.: Penicillinase and ampicillin resistance in a strain of *Escherichia coli*. J. gen. Microbiol. **30**, 299–306 (1963).
SMITH, J. T.: R-factor gene expression in gram-negative bacteria. J. gen. Microbiol. **55**, 109–120 (1969).
SMITH, J. T., HAMILTON-MILLER, J. M. T.: Differences between penicillinases from gram-positive and gram-negative bacteria. Nature (Lond.) **197**, 976–978 (1963).

SOMPOLINSKY, D., BEN-YAKOV, M., ABOUD, M., BOLDUR, I.: Transferable resistance factors with mutator effect in *Salmonella typhi*. Mutation Res. **4**, 119–127 (1967).

SPEYER, J. F., LENGYEL, P., BASILIO, C.: Ribosomal location of streptomycin sensitivity. Proc. nat. Acad. Sci. (Wash.) **48**, 684–686 (1962).

SPEYER, J. F., LENGYEL, P., BASILIO, C., WAHBA, A. J., GARDNER, R. S., OCHOA, S.: Synthetic polynucleotides and the amino acid code. Cold Spr. Harb. Symp. quant. Biol. **28**, 559–567 (1963).

SPOTTS, C. R.: Physiological and biochemical studies on streptomycin dependence in *Escherichia coli*. J. gen. Microbiol. **28**, 347–365 (1962).

SPOTTS, C. R., STANIER, R. Y.: Mechanism of streptomycin action on bacteria: a unitary hypothesis. Nature (Lond.) **192**, 633–637 (1961).

STAEHELIN, T., MESELSON, M.: Determination of streptomycin sensitivity by a subunit of the 30S ribosome of *Escherichia coli*. J. molec. Biol. **19**, 207–210 (1966).

SUAREZ, G., NATHANS, D.: Inhibition of aminoacyl-s-RNA binding to ribosomes by tetracycline. Biochem. biophys. Res. Commun. **18**, 743–750 (1965).

SUZUKI, I., KAJI, H., KAJI, A.: Binding of specific s-RNA to 30S ribosomal subunits: effects of 50S ribosomal subunits. Proc. nat. Acad. Sci. (Wash.) **55**, 1483–1490 (1966).

SUZUKI, Y., OKAMOTO, S., KONO, M.: Basis of chloramphenicol resistance in naturally isolated resistant *Staphylococci*. J. Bact. **92**, 798–799 (1966).

SWALLOW, D. L., SNEATH, P. H. A.: Studies on staphylococcal penicillinase. J. gen. Microbiol. **28**, 461–469 (1962).

SYPHERD, P. S., STRAUSS, N.: The role of RNA in repression of enzyme synthesis. Proc. nat. Acad. Sci. (Wash.) **50**, 1059–1065 (1963).

TAYLOR, A. L., TROTTER, C. D.: Revised linkage map of *Escherichia coli*. Bact. Rev. **31**, 332–353 (1967).

TRAUB, P., HOSOKAWA, K., NOMURA, M.: Streptomycin sensitivity and the structural components of the 30S ribosomes of *Escherichia coli*. J. molec. Biol. **19**, 211–214 (1966).

TRAUB, P., NOMURA, M.: Structure and function of *E. coli* ribosomes. V. Reconstitution of functionally active 30S ribosomal particles from RNA and proteins. Proc. nat. Acad. Sci. (Wash.) **59**, 777–784 (1968).

TRAUT, R. R.: Acrylamide gel electrophoresis of radioactive ribosomal protein. J. molec. Biol. **21**, 571–576 (1966).

TRAUT, R. R., MONRO, R. E.: The puromycin reaction and its relation to protein synthesis. J. molec. Biol. **10**, 63–72 (1964).

UMEZAWA, H., OKANISHI, M., KONDO, S., HAMANA, K., UTAHARA, R., MAEDA, K., MITSUHASHI, S.: Phosphorylative inactivation of aminoglycosidic antibiotics by *Escherichia coli* carrying R factor. Science **157**, 1559–1561 (1967).

UNOWSKY, J., RACHMEIER, M.: Mechanisms of antibiotic resistance determined by resistance-transfer factors. J. Bact. **92**, 358–365 (1966).

VAZQUEZ, D.: Uptake and binding of chloramphenicol by sensitive and resistant organisms. Nature (Lond.) **203**, 257–258 (1964a).

VAZQUEZ, D.: The binding of chloramphenicol by ribosomes from *Bacillus megaterium*. Biochem. biophys. Res. Commun. **15**, 464–468 (1964b).

VAZQUEZ, D.: Binding of chloramphenicol to ribosomes: the effect of a number of antibiotics. Biochim. biophys. Acta (Amst.) **114**, 277–288 (1966a).

VAZQUEZ, D.: Antibiotics affecting chloramphenicol uptake by bacteria: their effect on amino acid incorporation in a cell-free system. Biochim. biophys. Acta (Amst.) **114**, 289–295 (1966b).

VAZQUEZ, D.: Mode of action of chloramphenicol and related antibiotics. In: Biochemical studies of antimicrobial drugs. Symp. Soc. gen. Microbiol. **16**, 169–191 (1966c).

VAZQUEZ, D., MONRO, R. E.: Effects of some inhibitors of protein synthesis on the binding of aminoacyl-t-RNA to ribosomal subunits. Biochim. biophys. Acta (Amst.) **142**, 155–173 (1967).

WARING, M. J.: The effects of antimicrobial agents on ribonucleic acid polymerase. Molec. Pharmacol. **1**, 1–13 (1965).

WATANABE, T.: Infective heredity of bacterial drug resistance. Bact. Rev. **27**, 87–115 (1963).

WATANABE, T., FUKASAWA, T.: Episome-mediated transfer of drug resistance in *Enterobacteriaceae*. II. Elimination of resistance factors with acridine dyes. J. Bact. **81**, 679–683 (1961a).

WATANABE, T., FUKASAWA, T.: Episomic resistance factors in *Enterobacteriaceae*. XII. Chromosomal attachment of resistance transfer factor in *Escherichia coli* strain K12. Med. Biol. (Tokyo) **59**, 180–184 (1961b); quoted by WATANABE (1963).

WATANABE, T., FUKASAWA, T.: Episome-mediated transfer of drug resistance in *Enterobacteriaceae*. IV. Interactions between resistance transfer factor and F-factor in *Escherichia coli* K12. J. Bact. **83**, 727–735 (1962).

WEBER, K., OSBORN, M.: The reliability of molecular weight determinations by dodecyl sulfate-polyacrylamide gel electrophoresis. J. biol. Chem. **244**, 4406–4412 (1969).

WEISBERGER, A. S., WOLFE, S., ARMENTROUT, S.: Inhibition of protein synthesis in mammalian cell-free systems by chloramphenicol. J. exp. Med. **120**, 161–181 (1964).

WINSHELL, E., SHAW, W. V.: Kinetics of induction and purification of chloramphenicol acetyltransferase from chloramphenicol-resistant *Staphylococcus aureus*. J. Bact. **98**, 1248–1257 (1969).

WOLFE, A. D., HAHN, F. E.: Mode of action of chloramphenicol. IX. Effects of chloramphenicol upon a ribosomal amino acid polymerization system and its binding to bacterial ribosomes. Biochim. biophys. Acta (Amst.) **95**, 146–155 (1965).

WOLLMAN, E. L., JACOB, F.: Sur les processus de conjugaison et de recombinaison chez *E. coli*. II. La localisation chromosomique du prophage λ et les conséquences génétiques de l'induction zygotique. Ann. Inst. Pasteur **93**, 323–339 (1957).

YAMADA, T., TIPPER, D., DAVIES, J.: Enzymatic inactivation of streptomycin by R factor-resistant *Escherichia coli*. Nature (Lond.) **219**, 288–291 (1968).

YEE, R. B., GEZON, H. M.: Ribonucleic acid of chloramphenicol treated *Shigella flexneri*. J. gen. Microbiol. **32**, 299–306 (1963).

ZEEUW, J. R. DE: Accumulation of tetracyclines by *Escherichia coli*. J. Bact. **95**, 498–506 (1968).

Some Recent Developments in the Physiology of the Thyroid Gland

JOHN B. STANBURY*

With 3 Figures

Table of Contents

Thyroid endocrinology has often been revolutionized. Two generations ago, two events radically altered clinical thyroidology. These were the introduction of iodine for prophylaxis against endemic goiter and for preparation of the patient with Basedow's disease who was about to undergo surgical removal of the gland. The field surged forward again a generation later when within

*Unit of Experimental Medicine, Department of Nutrition and Food Science, Massachusetts Institute of Technology, Cambridge, Massachusetts 02139.

two years the antithyroid drugs were discovered and evidence was obtained that radioactive iodine could be used in the treatment of the overactive thyroid gland.

Recently, a revolution has moved the field ahead once again, and this time in several directions. Among the innovations are the discovery and characterization of the long-acting thyroid stimulator and the thyrotropin-releasing factor, identification of the central role of cyclic adenosine monophosphate in the intermediary metabolism of the thyroid, and appreciation of genetic and immunologic factors in thyroid disease. In this review, several of the new advances are examined to bring them into perspective with traditional and established views of thyroid structure and function. Choice of subject is the author's caprice; neglect of some important topics reflects the limitations of time and space.

I. Thyrotropic Hormone and the Role of Cyclic 3′,5′-Adenosine Monophosphate

The concept of an intimate functional relationship between the thyroid and the pituitary goes back at least to ROGOWITSCH (1888) who described cellular changes in the anterior lobe of the pituitary after removal of the thyroid gland in rabbits. Concepts of reciprocal or negative feedback control evolved from the observations of LOEB and KAPLAN (1924). They noted that the anterior lobe of the pituitary became hypertrophied after thyroidectomy, but that this could be prevented by feeding the animals desiccated thyroid. Over the succeeding years, an enormous effort was exerted toward pin-pointing the principle of the anterior pituitary involved in regulating thyroid function and toward characterizing its effects. Much has been achieved in this direction, but less attention has been accorded the interaction between the thyroid hormones and the responding cells of the anterior pituitary, since the relevant cells of the pituitary have not yet been isolated for individual physiological and biochemical scrutiny.

Thyrotropic hormone (TSH) has been obtained in highly purified form, and many of its chemical and physical properties have been determined. It is a mucoprotein with a molecular weight of approximately 26000, makes its appearance at the twelfth week of fetal life, and is synthesized and stored in basophilic Schiff-positive cells of the anterior pituitary as cytoplasmic granules. These granules may be solubilized by deoxycholate to yield a monomer of TSH (KRASS et al., 1969), which is digested by papain to yield a biologically active dialyzable component with a molecular weight of approximately 7000. Thus, TSH appears to contain a biologically active core surrounded by a non-essential, papain-digestible peptide.

A. Control of Synthesis and Secretion

The traditional model of thyroid control has a negative feedback loop between the TSH-secreting cells of the pituitary and the thyroid in order to maintain function at a preselected or set point. Actually, control is considerably more complex. Thus there are poorly defined control loops for the cells of the hypothalamus that secrete the TSH-releasing factor, a loop involving the amount of iodide in the thyroid gland itself, a loop involving direct control of the thyroid by thyroid hormone, and probably others. The level of operation of the system is governed by the equilibrium existing within a matrix of these several feedback loops. The "set point" is determined by the equilibrium state among the loops; a predetermined set point is not required. "Equilibrium point" would be a more suitable term.

The major factor controlling secretion of TSH is the plasma concentration of thyroxine. Destruction of the thyroid induces a 15-fold increase in the ^{3}H-thymidine labeling frequency of pituitary cells within 12 months, but the frequency returns to normal within 5 to 10 days after the start of thyroxine therapy (MESSIER, 1969). Incorporation of ^{3}H-uridine into RNA is also enhanced by thyroidectomy and depressed by administration of triiodothyronine (LEE et al., 1968). Changes in the opposite direction occur in the liver. Repletion of pituitary TSH after depletion by thiouracil treatment does not occur unless a small amount of thyroid hormone is present. This phenomenon could occur either because TSH is secreted under these circumstances as soon as it is synthesized or because a small amount of thyroid hormone is essential for TSH synthesis.

The pituitary-thyroid system is sluggish in comparison with most other endocrine systems, and synthesis of TSH is turned on much more slowly than it is turned off. When the plasma thyroid hormone was reduced rapidly by exchange transfusions, no increase in thyroid hormone secretion was observed for several hours, and 5 days were required for the free thyroxine level of the plasma to reach normal (SUEMATSU et al., 1969). By contrast, there is an almost immediate fall in plasma TSH upon administration of thyroid hormone.

TSH secretion has been studied in pituitary explants. Secretion was stimulated by dibutyryl cyclic 3′,5′-adenosine monophosphate (DB_cAMP) and by theophylline. The secretion was inhibited by thyroxine, enhanced by epinephrine and phentolamine, and partially blocked by propranolol (WILBER et al., 1969). The rat pituitary gland incorporates ^{14}C-labeled glucosamine into TSH *in vitro*; this, too, was enhanced by addition of propylthiouracil to the diet of the rats and was depressed when thyroxine was given. Incorporation was not increased by TSH-releasing factor (TRF), although TSH was released.

The normal pituitary has a functional heterogeneity with respect to secretion of TSH. Sinusoidal blood collected in the course of pituitary surgery varied in concentration of TSH from less than 0.4 ng to more than 2.6 μg/ml in samples from various regions of the gland (CONWAY et al., 1969). Human pituitary cells in explant culture produce immunoreactive TSH, but this secretion falls rapidly and may be undetectable after 4 months of subculture. Adenoma tissue in culture maintains a detectable level of TSH secretion (KOHLER et al., 1969).

Vasopressin, oxytocin, and epinephrine release TSH from anterior pituitary tissue and stimulate ^{14}C-glucose oxidation *in vitro* (KRASS et al., 1968). Vasopressin fails to stimulate the release of hornone from an exteriorized sheep-thyroid preparation (FALCONER, 1968) and is not active *in vitro* (READ et al., 1969). Oxidation of glucose is unaffected by all three hormones in concentrations that are maximally effective in releasing TSH (KRASS et al., 1968). The stimulatory effect of epinephrine but not of vasopressin on TSH release is blocked by phenoxybenzamine. From these results, no clear image emerges of the intimate biochemical control of TSH synthesis and secretion. Although it is clear that synthesis and secretion of TSH are dominated by the concentration of thyroid hormone in blood perfusing the pituitary, it is unclear how the effect is mediated. There is evidence that cyclic AMP and the adenyl cyclase system are involved. An important area of inquiry for the future will be the intermediary metabolic events between thyroid hormone influence on the pituitary cell and TSH synthesis and secretion.

B. The Effects of TSH on the Thyroid Gland

Thyrotropic hormone has many effects on the thyroid gland. The field has been critically reviewed recently (SCHELL-FREDERICK and DUMONT, 1970). Responses are seen in the metabolism of proteins, carbohydrates, lipids, and nucleic acids, as well as in the configuration of the cells and their growth and division. There have been two conflicting views accounting for these phenomena. One is that TSH has multiple direct effects on several different systems within the thyroid cell; the other is that TSH has a single site of action and that all of the observed effects are secondary, some being specific to the thyroid cell because of its own specialized internal differentiation. The dominant concept at present is the latter, as will become apparent when various aspects of TSH action are examined in the paragraphs that follow.

1. Glucose Metabolism. TSH stimulates the uptake of glucose by sheep, dog, and calf thyroid slices when high concentrations of the hormone are used. On the other hand, TSH decreases the incorporation of glucose into glycogen-like components of thyroid slices and rapidly elevates phosphorylase A levels. This latter effect may be due to increased tissue levels of 5′-AMP, which activates phosphorylase, rather than an increase in phosphorylase activity *per se*. The effect of TSH evidently is to diminish glycogen synthesis rather than to speed its degradation.

Formation of lactate is increased by TSH. This effect clearly is mediated by cyclic AMP, which activates phosphofructokinase, the rate-limiting enzyme in glycolysis. The pentose-phosphate pathway of glucose metabolism is inhibited by low concentrations of TSH and activated by higher concentrations. The stimulating effect may not be mediated by the cyclic nucleotide. Pyruvate oxidation to CO_2 is stimulated rapidly by TSH. This phenomenon, like the oxidation of glucose, occurs only at higher concentrations of TSH and probably reflects an activation of mitochondrial respiration.

2. Protein Synthesis and Degradation. Although TSH appears to increase the transport of amino acids into the thyroid cell, the rise in protein synthesis is not fully accounted for by the increase in amino acid pool sizes and actually represents a net stimulation of protein synthesis. The number of polyribosomes increases at the expense of monosomes as a result of incubation with TSH (DUMONT, 1968), thus providing additional evidence in favor of increased net protein synthesis. TSH has long been known to stimulate the number of colloid droplets in the thyroid epithelial cells. Careful inspection of the colloid surface of the cells shows an initial increase in pseudopod formation. These pseudopods entrap colloid, which then appears as droplets within the cell. Subsequently the droplets fuse with lysosomes, and digestion of the entrapped thyroglobulin ensues. Chlorpromazine, a lysosomal stabilizing substance, inhibits TSH stimulation of colloid-droplet fromation in the thyroid (ONAYA and SOLOMON, 1970), as does actinomycin (FUJITA and SUEMASA, 1968). Colloid-droplet formation is, however, also stimulated by DB_cAMP; this finding suggests that the TSH effect is mediated through the cyclic nucleotide.

Amino acid incorporation increases even after preincubation of tissue with actinomycin D. Since impaired glucose metabolism abolishes protein synthesis resulting from TSH stimulation, enhancement of protein synthesis may be secondary to stimulation of glucose metabolism (TONG, 1967). Cycloheximide and puromycin, which are inhibitors of protein synthesis, and actinomycin inhibit release of radioiodine from the thyroid, but a time lapse of 7 hours or more is required. Evidently, new protein synthesis is not required for the acute effects of TSH but is required for the late effects (MCKENZIE et al., 1968).

3. Nucleotide Synthesis. Synthesis of RNA is stimulated by TSH and by the long-acting thyroid stimulator (LATS) (OCHI and DEGROOT, 1968). TSH also stimulates RNA synthesis in cultures of fetal rat thyroids (IMBENOTTE et al., 1969) and incorporation of orotic acid into pyrimidine nucleotides (LINDSAY et al., 1969) but causes no noteworthy change in total pyrimidine nucleotides. It increases the oxidation of NADPH and NADH (ZAKARIJA et al., 1969).

TSH accelerates incorporation of ^{3}H-thymidine into thyroid cells and accelerates the rate of loss as well (SPEIGHT et al., 1968). In the chick embryo at about the eleventh day, when endogenous TSH secretion begins, there is a sharp drop in labeling of the nuclei. Evidently, TSH switches the embryonic thyroid gland from rapid cell reproduction to glandular function (STRAZNICKY and MESS, 1967).

4. Lipid Synthesis. Glucose, glycerol, palmitate, and oleate are incorporated into phosphatidylinositol under the influence of TSH. TSH also increases the incorporation of palmitate into 1,2-diglycerides but not incorporation into phosphatidylcholine, phosphatidylethanolamine or 1,3-diglycerides (SCOTT et al., 1968). TSH causes a coordinated increase in thyroid phospholipid, RNA, and the iodoprotein associated with the endoplasmic reticulum (KERKOF and TATA, 1969), and increases incorporation of ^{32}P into phospholipid at the step beyond the synthesis of ATP (SCHNEIDER, 1969).

5. Iodine Metabolism and the Thyroid Hormones. The initial effect of TSH on the thyroid is to diminish iodide trapping, an effect apparently caused by an increase in the rate at which iodide leaves the thyroid. Later, TSH increases the entry of iodide into the thyroid cells by activation of transport and also rapidly stimulates organification of iodine. TSH increases the ratio of monoiodotyrosine to diiodotyrosine and also the ratio of triiodothyronine to thyroxine but these effects depend also on the amount of iodine in the thyroid and are related more to iodine than to TSH (GREER et al., 1968; GREER and ROCKIE, 1969). Thyroglobulin can be formed in the absence of TSH but has a slightly lower sedimentation constant than normal and is unusually sensitive to degradation into half molecules in solutions of low ionic strength or high pH. This thyroglobulin is low in iodine (ROSENBERG and CAVALIERI, 1969; CAVALIERI et al., 1970).

6. Blood Flow. Blood flow in the thyroid is increased by TSH within minutes after injection. At the same time, the serotonin content of the gland falls (CLAYTON and SZEGO, 1967), but this effect does not extend to mast cells of extrathyroidal tissues (CLAYTON and MASUOKA, 1968). The number of mast cells in the thyroid is increased by TSH, but there is no certainty that TSH stimulates mast cells in the thyroid directly. This phenomenon could well be mediated by a factor or factors escaping from the contiguous stimulated thyroid cells.

7. Other TSH Effects. The response of virtually every biochemical process of the thyroid has been studied in terms of its response to TSH, and the effects of many biochemical inhibitors and cell stimulants have been studied with respect to TSH effect. It is not possible to review them all here. The reader may refer to the excellent review of DUMONT et al. (1969). Recent additions to the catalogue of effects include a demonstration that the responsiveness of the thyroid to TSH is reduced by hypoxia. Also, dinitrophenol blocks the growth response of the thyroid to TSH but does not block glandular iodine metabolism (MAAYAN, 1968). The tyrosine content of the thyroid is increased, as is its concentration with respect to tissue RNA; this rise is

enhanced by administration of propylthiouracil (HODGE et al., 1969). TSH increases uptake of ^{35}S sulfate by the harderian gland of the mouse, but serum containing a high titer of LATS has no such effect (SINGH and MCKENZIE, 1969).

C. Mediation of TSH through Cyclic AMP

It is now well established that cyclic AMP is produced in many cells as a result of hormone effects. In many instances, the effect of TSH is channeled through cyclic AMP which acts as a "second messenger" (Table 1). TSH affects a wide spectrum of activities of the thyroid cell. There has been a

Table 1. Partial list of hormones which exert their effect through stimulation of adenyl cyclase and increase of cyclic AMP

Hormone	Target	Effect
TSH	thyroid epithelial cell	increased hormone synthesis and release
ACTH	adrenal cortex	steroidogenesis
LH	corpus luteum	steroidogenesis
LH	Leydig cells	steroidogenesis
Parathormone	renal tubule	phosphate resorption
Glucagon	β cell of pancreas	insulin release
Glucagon	liver cell	glycogenolysis
Epinephrine	liver cell	glycogenolysis
Epinephrine	skeletal muscle	phosphorylase activation
Epinephrine	fat cell	lipolysis
MSH	melanocyte	melanocyte dispersion
Gastrin	stomach	gastric acid secretion

recent exhaustive review of the evidence bearing on the problem of whether these effects are all channeled through an interaction with the thyroid plasma membrane, which activates adenyl cyclase, or whether there are other effects of TSH not mediated through this route (SCHELL-FREDERICK and DUMONT, 1970). Much evidence supports the view that most, if not all, of the activity of TSH is exerted through cyclic AMP, but in many instances the mediation has not been firmly established. The cyclic nucleotide increases 1-^{14}C-glucose metabolism within 15 minutes. Theophylline, which impairs diesterase degradation of cyclic AMP, also stimulates glucose metabolism and is more than additive with cyclic AMP. Cyclic AMP is effective in a concentration as low as 8×10^{-4} M (MACCHIA et al., 1969).

Release of thyroid hormone from mouse thyroid *in vivo* is increased by cyclic AMP and is augmented by simultaneous administration of theophylline. OCHI and DEGROOT (1969a) found that release *in vivo* but not *in vitro* is also effected by 5′-AMP, 5′-ADP, and 5′-ATP.

The cyclic nucleotide promotes release of ^{131}I bound *in vivo* and increases ^{131}I labeling of thyronines in mouse thyroids *in vitro*; these effects are potentiated by theophylline (ENSOR and MUNRO, 1969). DB_cAMP stimulates secretion of hormonal ^{131}I and inorganic iodide from the dog thyroid within 20 to 30 minutes (AHN et al., 1969). Colloid droplets appear in the thyroid parenchymal cells of the dog but not the rat after administration of DB_cAMP (PASTAN and WOLLMAN, 1967).

Triiodothyronine and thyroxine are released *in vitro* by DB_cAMP (TONOUE et al., 1970). Proteolysis of thyroglobulin is stimulated both by DB_cAMP and by TSH, but the stimulating effects at maximally effective concentrations of each are not additive. Thus, a common pathway of action appears to be involved (AHN and ROSENBERG, 1970a). TSH and DB_cAMP have essentially identical effects on iodine incorporation into thyroglobulin in suspended bovine thyroid cells. Incorporation of labeled leucine into protein is also stimulated (WILSON et al., 1968).

The concentration of cyclic AMP (GILMAN and RALL, 1968a) and the activity of adenyl cyclase (KANEKO et al., 1969) are increased in the thyroid within 1 minute by TSH. The increase in AMP concentration, which reaches a maximum in 3 to 6 minutes (GILMAN and RALL, 1968a), parallels the concentration of TSH over a wide range.

LISSITZKY et al. (1969) found that cyclic AMP increased protein synthesis in a cell-free system using thyroid polyribosomes, but the system was not stimulated by TSH. This finding was consistent with cyclase activation at the plasma membrane mediating the TSH effect.

Organic binding of iodine and formation of thyroxine are stimulated by TSH, DB_cAMP, fluoride, and prostaglandin E_1. Both TSH and fluoride activate adenyl cyclase in sheep thyroid slices, and either one increases the activity in the presence of a maximally stimulating quantity of the other. This effect indicates differences in the sites of action of these compounds. The fluoride effect is not stopped by adrenergic blocking drugs that inhibit TSH effects on cyclase (BURKE, 1970a). TSH and prostaglandin E_1, but not fluoride, enhance synthesis of cyclic AMP in dog thyroid slices (AHN and ROSENBERG, 1970b). Stimulation of organic binding of iodine by TSH and prostaglandin E_1 thus appears to be mediated through increased formation of cyclic AMP, but fluoride acts through a different mechanism. Fluoride, which stimulates thyroid adenyl cyclase activity in homogenates, fails to increase the concentration of cyclic AMP in canine thyroid slices as noted above. Two other substances, carbamylcholine and menadiol sodium diphosphate, increase glucose oxidation in thyroid slices but do not change the concentration of cyclic AMP (KANEKO et al., 1969). The increased cyclase activity is transient, but increased glucose metabolism persists for as long as 1 hour. TSH does not stimulate adenyl cyclase activity in other glands. Resembling TSH, prostaglandin E_2 stimulates glucose oxidation but fails to increase cyclic AMP levels. Prostaglandin E_1 changes neither. These findings are consistent with the concept that the effects of TSH are channeled through cyclic AMP but that fluoride and other substances may influence gland metabolism through different pathways (GILMAN and RALL, 1968a; ZOR et al., 1969).

WOLFF et al. (1970) found that stimulation of adenyl cyclase activity by TSH in a membrane preparation from bovine thyroid was inhibited by low concentrations of lithium ions. Lithium inhibits several aspects of iodine metabolism in the thyroid (HULLIN and JOHNSON, 1970), possibly through rate-limiting of adenyl cyclase.

D. Alternative Routing of TSH Effects

From the foregoing, it appears that many responses of the thyroid to TSH are mediated through adenyl cyclase and cyclic AMP, but TSH may also have other effects on the thyroid. Much of the evidence is pharmacological and must be interpreted with caution because of dosage effects and the non-specificity and ambiguity sometimes associated with use of agents such as α- and β-adrenoceptor-blocking drugs, xanthines, and others (BURKE, 1969a; LEVEY et al., 1969). From evidence derived from studies with adrenoceptor-blocking agents and catecholamines, it appears that TSH stimulation of adenyl cyclase does not involve an adrenoceptor mechanism (BURKE, 1969a).

BURKE (1969b, 1970b) has found concentrations of TSH that increase phospholipogenesis but have no effect on adenyl cyclase activity or endocytosis. On the other hand, prostaglandin E_1 stimulates adenyl cyclase, endocytosis, and glucose oxidation without affecting phospholipogenesis. Thus, cyclase activation and stimulation of endocytosis may be a result of stimulation of glucose oxidation, but TSH activation of phospholipogenesis may be mediated in another manner.

Several other observations are difficult to explain by the hypothesis that cyclic AMP mediates the actions of TSH. Prostaglandin E_1 increases the intracellular levels of cyclic AMP and stimulates glucose oxidation and iodide binding to protein, but neither fluoride, which activates adenyl cyclase in cell-free systems, nor prostaglandin E_1 has any effect on intracellular colloid droplet formation, an early effect of TSH action (RODESCH et al., 1969). On the other hand, BURKE (1970b) found an effect of prostaglandin E_1 and endocytosis, and DUMONT and his colleague (WILLEMS et al., 1969) observed increased endocytosis, glucose metabolism, and covalent binding of iodide to protein in dog thyroid slices in the presence of TSH or cyclic AMP; the effects were increased by caffeine. KERKOF and TATA (1969) found that the rapid effects of TSH on transport and metabolic functions were mimicked by DB_cAMP but that the latter failed to influence the slower biosynthetic responses.

BURKE (1969c) has compared the effects of TSH and DB_cAMP on the metabolism of 1-^{14}C-glucose and pyridine nucleotide. DB_cAMP regularly stimulated glucose oxidation and phospholipogenesis, and decreased the ratios of oxidized to reduced pyridine nucleotides, whereas there was little correlation between TSH stimulation of glucose oxidation, phospholipogenesis, and changes in pyridine nucleotide concentration, or ratios of oxidized to reduced forms. Thus, there appeared to be a difference between the effects of the peptide stimulator and the stimulation from cyclic AMP.

BURKE (1969b, 1970b) has also questioned the role of cyclic AMP in TSH mediation. He observed no potentiation of TSH by cyclic AMP either on glucose oxidation or phospholipogenesis and also found no interaction between theophylline and submaximal doses of TSH on either glucose oxidation or ^{32}P incorporation. He concluded that at least some of the effects of TSH on thyroid intermediary metabolism are not mediated by cyclic AMP. GILMAN and RALL (1968b) also found that cyclic AMP did not appear to be the mediator of TSH-stimulated glucose oxidation.

Activation of adenyl cyclase by TSH is impaired by three adrenoceptor-blocking agents, DL-propranolol, D-propranolol, and phentolamine, in concentrations approximately 100 times those required to produce α- and β-blockade. A concentration of DL-propranolol and phentolamine that impairs TSH stimulation of glucose oxidation in dog thyroid slices does not impair stimulation of glucose oxidation by DB_cAMP. TSH still stimulates glucose oxidation in bovine thyroids at a concentration of propranolol that abolishes TSH stimulation of adenyl cyclase (LEVEY et al., 1969). Chlorpromazine, a drug that stabilizes lysosomes, and propranolol block the effect of TSH on endocytosis. These drugs also block stimulation of glucose oxidation by TSH. Propranolol also inhibits the effect of DB_cAMP. This indicates that propranolol blocks beyond the TSH effect on adenyl cyclase. Both drugs appear to stabilize the lysosomal membranes and other membranes as well.

E. Scheme of Action

From the foregoing it is well established that TSH activates the cyclase system of the plasma membrane and that this is followed by the appearance of an increased concentration of cyclic AMP in the thyroid cell. Secondary effects then occur, including glucose oxidation, colloid droplet formation, phosphorus incorporation into phospholipid, protein synthesis, and enhanced iodine metabolism. A current interpretation is shown diagrammatically in Fig. 1 (from SCHELL-FREDERICK and DUMONT, 1970). After careful review of the available information, the authors concluded that "for only one TSH effect, i.e., the stimulation of the pentose-phosphate pathway by high concentrations of hormone, are there indications that cyclic AMP may not be involved; whether this hormonal effect is of physiological significance is doubtful". They concluded that the available evidence is compatible with the hypothesis that all other effects of TSH are mediated by cyclic AMP; also

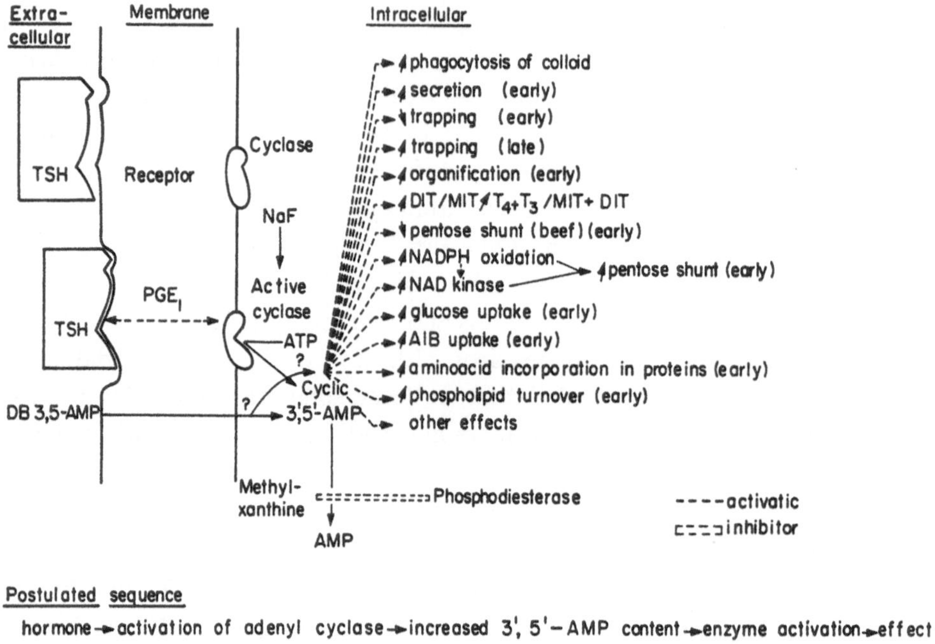

Extra-cellular

Membrane

Intracellular

↑ phagocytosis of colloid
↑ secretion (early)
↓ trapping (early)
↑ trapping (late)
↑ organification (early)
↑ DIT/MIT ↑ T_4+T_3 /MIT+ DIT
↓ pentose shunt (beef) (early)
↑ NADPH oxidation → ↑ pentose shunt (early)
↑ NAD kinase
↑ glucose uptake (early)
↑ AIB uptake (early)
↑ aminoacid incorporation in proteins (early)
↑ phospholipid turnover (early)
other effects

TSH
Receptor
Cyclase
NaF
TSH
PGE_1
Active cyclase
ATP
?
Cyclic 3',5'-AMP
DB 3,5-AMP
?
Methyl-xanthine
Phosphodiesterase
AMP

---- activatic
⊏⊐ inhibitor

Postulated sequence

hormone → activation of adenyl cyclase → increased 3', 5'-AMP content → enzyme activation → effect

Fig. 1. The multiple effects of thyrotropic hormone on the thyroid cell. (From SCHELL-FREDERICK and DUMONT, 1970, reprinted by permission)

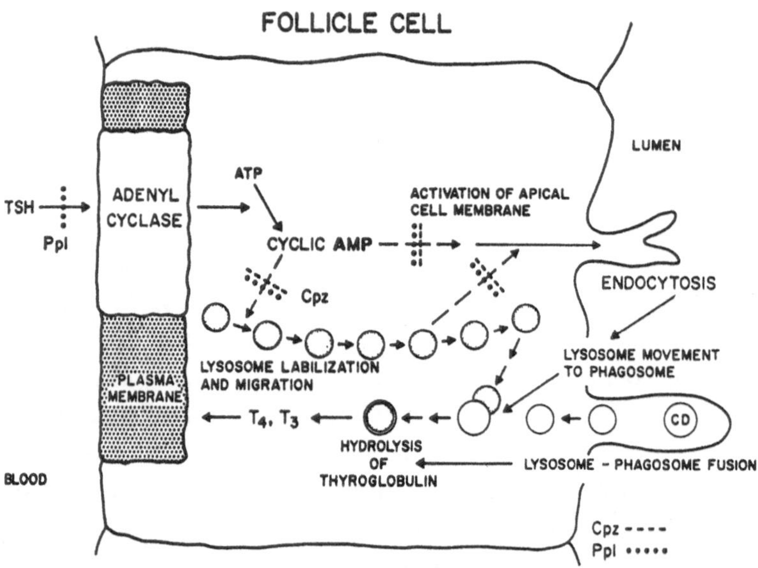

Fig. 2. Schema of steps in the secretion of thyroid hormones. Two alternate pathways are shown between cyclic AMP and the activation of endocytosis. Possible sites of inhibition of chlorpromazine (*Cpz*) and propranolol (*Ppl*) are indicated by the dashed and dotted lines respectively. (From ONAYA and SOLOMON, 1969, reprinted by permission)

that some of the TSH effects, particularly the delayed hormonal effects, deserve further study in this regard.

With this as background, ONAYA and SOLOMON (1969) have proposed a scheme in which TSH activates adenyl cyclase in the plasma membrane (Fig. 2). The resulting cyclic AMP travels to the luminal side of the thyroid cell and activates the apical cell membrane to undertake colloid endocytosis. Lysosomes labilized by cyclic AMP fuse with the colloid droplets, whereupon the thyroglobulin undergoes hydrolysis and releases thyroid hormone into the blood. This model has much to recommend it. Studies with detergents and chlorpromazine have suggested that these exert their effects on hormone release by effects on stabilizing and destabilizing the apical cell membrane (WILLIAMS and WOLFF, 1971).

Perhaps the time has now arrived when less attention need be paid to minute discrepancies between effects of TSH and DB_cAMP and more to precisely what these substances do in enzyme activation and inhibition. One would like to know, for example, whether stimulation of the thyroid cell is heralded by activation of a kinase involved in thyroid hormone synthesis. Does the nuclear chromatin of the thyroid cell respond in a specific mode to nonspecific activation by adenyl cyclase in order that synthesis of thyroglobulin may occur? The thyroid system should provide a unique opportunity for the study of cell specificity.

II. The Thyroid Cell Surface

The bounding structure of the thyroid parenchymal cell is the familiar double-layered plasma membrane. At the outer surface of the cell this membrane is in contact with the follicular basement membrane. Laterally, the cell touches its nearest neighbors, and there the adjacent membranes occasionally fuse to form desmosomes. Internally, the plasma membrane has long folds or projections, the microvilli, which are in intimate contact with the follicular colloid. Processes of colloid ingestion and the iodination reaction take place at this interface. Uniqueness of the microchemical structure of this membrane may be fundamental in the peculiar cytoarchitecture of the thyroid and in the specialized function of the thyroid cell in responding to TSH.

There is abundant evidence now that most of the adenyl cyclase activity of cells occurs in the plasma membrane. Also, as has already been cited, most if not all the effects of TSH are mediated through enhancement of cyclase activity and synthesis of cyclic AMP. Logically one might conclude that the effect of TSH is exerted at the plasma membrane, where it induces activation of adenyl cyclase. This hypothesis appears to be substantiated by additional information.

PASTAN et al. (1966) first demonstrated that thyroid cells exposed to TSH and then washed showed a persistence of hormone effect, but if these cells were exposed to anti-TSH antibody or to trypsin, the effect was obliterated. The TSH effect employed in these experiments was the easily measured and ready response of the thyroid cell in metabolizing 1-^{14}C-glucose. The authors pointed out that the antibody used in the experiments had a molecular weight of approximately 150000 and accordingly must have been effective on the TSH entrapped on the cell surface. Furthermore, trypsin itself did little to alter the metabolism of glucose by the cells or their ability to respond to TSH after it had been washed away. Again, the evidence favored a TSH action at the cell surface. Similar findings have been reported with adrenal and fat cells in relation to ACTH (TAUNTON et al., 1967). Corroborating evidence has been that ACTH, firmly bound to a large cellulose resin which could not pass the plasma membrane, is effective in stimulating the adrenal cell (SCHIMMER et al., 1968). Microscopically visible rounding up of adrenal cells when they were stimulated was cited as evidence of conformational change in the plasma membrane in response to the hormone.

Exposure of the thyroid cell to lecithinase C abolishes the response to TSH stimulation of both 1-^{14}C-glucose metabolism and phosphate incorporation into phospholipid, without affecting basal metabolic activity or the response to DB_cAMP (BURKE, 1969d; MACCHIA et al., 1970). The exposure reduced the lecithin content by about 50%. When thyroid slices were treated similarly with purified sphygomyelinase there was no blocking (MACCHIA et al., 1970). Neither enzyme blocked stimulation by DB_cAMP. The effect of lecithinase is exerted on both TSH and LATS stimulation (MACCHIA et al., 1970) but not on acetylcholine stimulation (BURKE, 1969d). There appears to be competitive interaction between LATS and TSH for binding sites at the thyroid cell surface (BURKE, 1969d). The thyroid of the patient with a high plasma titer of LATS (ADAMS et al., 1969) is not stimulated by TSH. Prior administration of TSH in mice impairs subsequent response to LATS and vice versa (BURKE, 1968b). Binding of TSH is affected directly by thyroxine; in this way thyroxine may regulate the biological activity of TSH directly (JACQUEMIN and HAYE, 1968).

Sodium ion is required for TSH action. Calcium ion is also required for maximal TSH action, but it is not established whether this is an effect solely on adenyl cyclase activity or is also related to TSH binding to the thyroid cell surface (ZOR et al., 1968). Ouabain, which impairs the ion-activated ATPase activity, abolished the effect of both LATS and TSH on phospholipogenesis (BURKE, 1969d).

The electrical properties of thyroid membranes have been studied by WILLIAMS (1970). Endogenous production of TSH was stimulated by administration of propylthiouracil or by cold stress. Both measures induced a fall in transmembrane potential. Rabbits pretreated with propylthiouracil had a higher capacitance than controls, possibly related to a larger surface membrane area. Addition of TSH to the perfusion fluid of thyroids *in vitro* also caused a rapid fall in resistance and increase in capacitance.

The effects of stimulating hormones in "ghosts" of responsive cells have been studied. The cells are prepared by exposure to hypotonic solution. The resulting preparations contain elements with nuclei, and most of the ghosts contain undefined particles and vesicles, but presumably these fractions are enriched in the plasma-membrane fraction. RODBELL (1967) found that a preparation from fat cells responded to ACTH with an increase in adenyl cyclase activity. Thyroid cell ghosts prepared by MAAYAN et al. (1970) retained a remarkably large repertoire of responses to TSH in metabolizing iodine but had lost the ability to increase glucose metabolism. A tenfold increase in the specific activity of 5′-nucleotidase activity indicated the preferential retention of plasma-membrane components. These results do not imply that iodine metabolism is a property of the plasma membrane.

The transport of iodide into the thyroid cell occurs against an electrochemical gradient and requires an expenditure of energy. The process is closely related to the sodium-potassium-ion-dependent ATPase activity of the cell, since it is inhibited by absence of either ion and is blocked by concentrations of ouabain which also block the sodium-potassium-ion-dependent ATPase. It has already been noted that many of the responses of the thyroid to TSH are inhibited by absence of sodium ions and by the presence of ouabain. These findings may be taken as further evidence that the TSH effect is exerted on the plasma membrane, since there is abundant evidence that the sodium- and potassium-ion-ATPase activity resides largely in the surface membrane of the cell.

Examination of the TSH-responsive components of thyroid cells has been difficult because of the peculiar structure of the gland and the lack of methods for identifying the precise origin of components of the differential separations. Whole homogenates of thyroid glands and particulate preparations respond to TSH with an increase in adenyl-cyclase activity (PASTAN and KATZEN, 1967; ZOR et al., 1969; YAMASHITA and FIELD, 1970; STANBURY et al., unpublished observations). YAMASHITA and FIELD (1970) have prepared membranes on gradients that had sodium- and potassium-ion-ATPase-specific activity ten times higher than that of the crude homogenate. Electron microscopy showed that the fraction contained minimal contaminating elements. An active membrane fraction has also been obtained by density gradient centrifugation (WOLFF, 1970). Adenyl cyclase was purified 60- to 140-fold and was stimulated by TSH. Membranes prepared in the author's laboratory (STANBURY et al., 1969) also have shown an increase in sodium- and potassium-ion-ATPase activity and a relative fall in NADH diaphorase activity in a fraction thought to be derived from the endoplasmic reticulum. 5′-Nucleotidase was much higher in the membranes that were also enriched in ATPase activity.

The method involved disruption by nitrogen cavitation and separation of membraneous components on discontinuous density gradients. Electron microscopy and scanning electron microscopy disclosed that these were spherical membraneous vesicles 100 to 200 nm in diameter. Approximately 15 protein components have been displayed on polyacrylamide gel gradients.

Study of the plasma membrane of the thyroid cell is in its infancy. Further work should provide insights into the nature of neoplastic change in these cells, the structure of TSH binding sites, clues to the remarkable microarchitecture of the thyroid and possibly clues to the nature of some thyroid diseases.

III. Thyrotropin-Releasing Factor

There has long been evidence that secretion of TSH is at least partially under hypothalamic control. Recently, the thyrotropin-releasing factor (TRF) secreted by the hypothalamus has been identified as a tripeptide, pyroglutamyl-histidyl-prolinamide (Fig. 3) (BOWERS et al., 1968a, 1968b, 1968c; FOLKERS et al., 1969; BOWERS et al., 1970; BURGUS et al., 1970; VALE et al., 1970). TRF is effective when administered directly into the pituitary (AVERILL, 1969a), when injected intravenously, or when ingested (SCHALLY, 1969), and is resistant to proteolytic activity (SCHALLY, 1969). It has been identified in the hypophyseal portal blood of the rat (WILBER and PORTER, 1970).

Administration of TRF causes an immediate release of TSH from the pituitary. This effect is not blocked by puromycin or cycloheximide and, accordingly, one may conclude that protein synthesis is not involved in this effect of TRF. The release of TSH brought about by TRF is blocked by triiodothyronine, and this effect of triiodothyronine is blocked by inhibitors of protein synthesis. Likewise, actinomycin D does not block the release of TRF but does block the inhibiting effect of triiodothyronine (BOWERS et al., 1968b).

TRF increases both synthesis and release of TSH in rat pituitary *in vitro* (MITTLER et al., 1969). It has been detected in the blood of thyroidectomized-hypophysectomized rats only after stimulation by cold (REDDING and SCHALLY, 1969). Synthesis of TSH is not stimulated acutely by TRF (WILBER and UTIGER, 1969a), whereas release of TSH by the rat pituitary *in vitro* is stimulated. Incorporation of labeled glucosamine into TSH is unchanged. However, pituitary glands removed from hypothyroid rats show increased incorporation of labeled glucosamine into TSH.

Injection of TRF into man causes a rise in plasma TSH concentration within 1 minute, a peak elevation at 30 minutes, and a return to normal levels within about 3 hours (BOWERS et al., 1968b; BAUGH et al., 1970; FLEISCHER et al., 1970). Its precise physiological role remains undetermined, and little information is currently available regarding the control of its secretion rate. The pituitary response to releasing factor is inhibited by thyroxine in

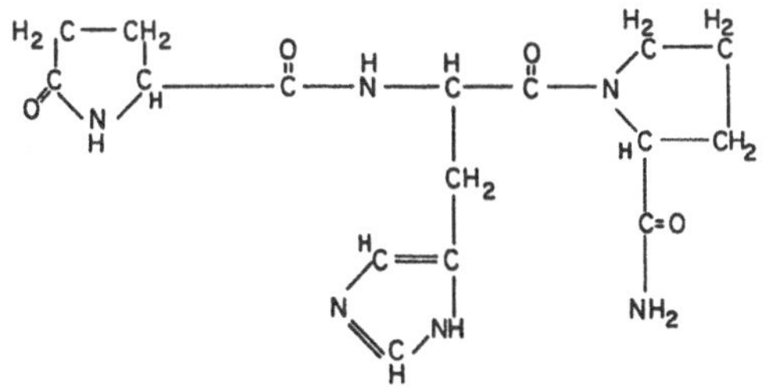

pyroglutamyl histidyl proline amide

Fig. 3. Structural formula of the thyrotropin-releasing factor

small doses (AVERILL, 1969b). Secretion of TSH normally is depressed by glucocorticoids, but these do not impair TRF-mediated secretion of TSH. These findings suggest that the glucocorticoid effect on TSH secretion is at a suprahypophyseal level (WILBER and UTIGER, 1969b). The thyroid effect on TSH secretion appears, on the contrary, to be at least primarily at the level of the pituitary, since in the absence of the hypophysis, the rat thyroid is exquisitely sensitive to the inhibiting effect of thyroxine on TSH plasma levels (MARTIN et al., 1970).

One may surmise that control of TSH release is complex, and is controlled both from higher neural centers and by the concentration of plasma thyroid hormone. Surely much remains to be learned regarding the details of the control of TSH secretion rate and the physiological and pathophysiological roles of TRF. A direct assay method for detecting the concentration of TRF in biological fluids is much needed. This will be a difficult problem to solve because TRF is secreted into the hypophyseal portal system and probably reaches the peripheral circulation in small amounts, if at all; even if it does reach peripheral circulation, its concentration there would not necessarily reflect accurately the events at the hypophyseal level. TRF involvement in diseases of the thyroid system remains unestablished.

IV. The Long Acting Thyroid Stimulator (LATS)

ADAMS (1958) first detected, in the plasma of patients with Graves' disease, a factor that had unusual properties in an assay test system. When injected, it caused a maximal stimulation of thyroid secretion at approximately 8 hours, whereas TSH reaches its peak effect within 1 or 2 hours. This new factor has now been studied extensively. It is found only in the plasma of patients who have or who have had Graves' disease, but is by no means detectable in all of them. In some patients, it reaches exceedingly high titer, especially in those with pretibial myxedema. It is a 7S gamma globulin of the I_gG class (BURKE, 1968a). It is secreted into the suspending fluid of cultured lymphocytes from patients with Graves' disease when the cells are stimulated with phytohemagglutinin (MCKENZIE and GORDON, 1965; MIYAI et al., 1967).

A. TSH and LATS

LATS differs from TSH in some of its physical properties. TSH can be extracted from serum both by ethanol-salt percolation and by fractional precipitation with acetone, while LATS is not extracted by either procedure (MCKENZIE, 1961). TSH is more resistant than LATS to heat inactivation and has a much shorter biological half-life when injected (ADAMS, 1960). Furthermore, the two stimulators have different sedimentation coefficients; TSH is recovered with the 4S fraction, whereas LATS is a 7S component. Finally, TSH but not LATS activity is inhibited by incubation with anti-TSH serum (ADAMS et al., 1962).

B. Effects on the Thyroid

Studies have failed to establish any essential difference between the effects of TSH and of LATS on the thyroid gland. Injection of LATS for several days causes increased weight of the gland, increased protein content, increased incorporation of tritiated leucine, and increase in the specific activity of protein-bound iodine. In chronic administration LATS does not appear to damage the thyroid cell in any way (OCHI and DEGROOT, 1969b). Release of thyroid radioiodine is stimulated by LATS. This release is blocked by actinomycin D, cycloheximide, and puromycin, but a time lapse of 8 hours or more is required for the inhibition to take effect (MCKENZIE et al., 1968).

The effect of LATS on ^{131}I uptake parallels the effects on glucose oxidation and phospholipogenesis in the thyroid. There is no correlation, however, between these changes and changes in pyridine nucleotide ratios; and the effects of the two peptide stimulators do not correlate with the effects of cyclic AMP on pyridine nucleotide levels (BURKE, 1969a). LATS and TSH also cause parallel and comparable rises in RNA synthesis and phospholipid formation (OCHI and DEGROOT, 1968). LATS interacts with the thyroid cell somewhat more slowly than does TSH (ADAMS, 1960). Studies comparing LATS and TSH influence on glucose oxidation and endocytosis suggest that these substances resemble one another qualitatively and that the action of LATS, if not slightly more than, is at least as rapid as that of TSH (SHISHIBA et al., 1970).

Both TSH and LATS stimulate proteolysis in the mouse thyroid gland, but apparently by making more substrate available rather than by early stimulation of proteolysis *per se* (KONNO et al., 1970). Both stimulate lipolysis in the rat epididymal fat cell (HART and MCKENZIE, 1971).

C. LATS as Immunoglobulin

It is well established that LATS is an immunoglobulin, but the problem of the protein against which it is directed is not settled. KRISS (1968) found that, in several differrent sera containing LATS, the activity was partially inhibited by anti-kappa antisera against the kappa and lambda immunoglobulin chains and that complete inactivation was achieved when both antisera were employed. This finding indicates that LATS is not derived from a monoclonal line of cells. DEAE Sephadex chromatography of sera containing LATS has resolved it into two zones in the I_gG globulin region (SMITH et al., 1969). It has been partially purified by adsorption onto and elution from thyroid microsomes and exhibited a high degree of affinity and specificity for the microsomes. LATS is a rather small part of the total I_gG population (WONG and LITMAN, 1969).

Whole homogenates of thyroid neutralize or absorb LATS activity. Most of this absorbing capacity appears to be in the soluble fraction containing several components including thyroglobulin, hemoglobin, and other serum proteins, but most of the absorption is by a 4S fraction (SMITH, 1970). BEALL et al. (1969a) found that LATS is inhibited by incubation with either particulate or soluble fractions. The inhibitor is widely distributed along a sucrose gradient of the soluble fraction of the homogenate.

The thyroid protein that acts as antigen is not known. BEALL et al. (1969b) immunized baboons with the microsomal fraction from human thyroids. The animals developed a rise in plasma TSH-like activity, whereas rabbits similarly immunized developed a LATS-like thyroid-stimulating substance. Injury of the thyroid with radioiodine therapy fails to induce the rise in LATS titer that might be expected if the injury caused a leak of a specific antigen into the blood (BURKE and SILVERSTEIN, 1969).

Thyroid-stimulating activity has been found in three different laboratories in the blood of rabbits injected with either thyroid homogenates or thyroid microsomes. No clinical sign of thyroid hyperfunction was detected (PINCHERA et al., 1965; MCKENZIE, 1968; SOLOMON and BEALL, 1968), but MCKENZIE (1968) found that the thyroids were not suppressible with triiodothyronine; presumably the stimulator was not TSH. More recently, SOLOMON and BEALL (1970) induced a thyroid-stimulating substance in rabbits by repeated immunization with human thyroid microsomes. This substance appeared to be of extrahypophyseal origin, was an immunoglobulin of the I_gG class, gave a response in the assay intermediate between those of TSH and LATS, was partially neutralized by antibovine TSH, and was not suppressed when plasma thyroxine concentration was above normal. The component was not TSH bound to a plasma globulin. Weight loss suggested that the rabbits may have been thyrotoxic.

D. LATS and the Cell Surface

Phospholipase C abolishes the response of thyroid slices to both TSH and LATS. This observation indicates that a phospholipid is required in the plasma membrane for response to the peptide hormones. Omission of sodium ions from the buffer or the presence of ouabain reduces the response to both peptide stimulators. It is worth noting that these two peptide hormones with strikingly different physical, chemical, and immunological properties appear to mediate in similar fashion the stimulatory action on the thyroid cell through interaction with a component or components of the thyroid cell membrane (Burke, 1969d).

E. LATS and the Cause of Graves' Disease

The initial detection of a factor in the blood of patients with Graves' disease that stimulated the thyroid, and that was different from TSH, promised new insight into the nature of the disease. The cause of the disorder has defied detection since its description in the first half of the nineteenth century. Increased secretion of thyrotropin was long thought to be involved, but improved methods for its detection in blood had failed to provide evidence that the over-active gland is associated with an elevated concentration of TSH in the plasma. Furthermore, there was never any good reason for suspecting that the exophthalmos and dermal changes of Graves' disease were a result of TSH. Something new was needed, and LATS seemed to provide it.

New disclosures suggested a close relationship, but not necessarily a causal one, between LATS and Graves' disease; among these were the high percentage of patients with circulating LATS (especially when concentrated gamma globulin was used for the assay) (Carneiro et al., 1966a), the correlation between the level of circulating LATS and the ^{131}I turnover rate (Carneiro et al., 1966b), and the temporary presence of circulating LATS in neonatal thyrotoxicosis (McKenzie, 1964; Sunshine et al., 1965), a transient disorder occurring in children of thyrotoxic mothers. Additional evidence was the finding from two laboratories that LATS was produced in the supporting medium containing lymphocytes from patients with Graves' disease but not from control subjects (McKenzie and Gordon, 1965; Miyai et al., 1967). Thus it was suggested that Graves' disease might arise through some error of the immune system that led lymphocytes to produce LATS. This hypothesis did not account for the ophthalmopathy or for the dermal changes. The former was perhaps attributable to an excess of an ophthalmotropic factor produced by the anterior pituitary and related in some undefined way to LATS.

Recent findings raise considerable doubt about a causal relationship between LATS and Graves' disease. Chopra et al. (1970) have developed a method for tenfold concentration of LATS from the serum. They have found that false positives occasionally appear when undiluted serum is used, but not when the serum is concentrated, and also that only 62% of concentrates from patients with active Graves' disease are positive for LATS. Thus, a large fraction of patients with typical Graves' disease do not have this circulating factor. CHOPRA and SOLOMON (1970) have also cited a patient whose course has provided unusually relevant information. LATS was detected in the plasma in high levels long before Graves' disease appeared. Later, at the time when the patient became clinically thyrotoxic and had elevated plasma concentrations of thyroid hormone, the LATS concentration in the plasma fell. Their studies on this patient have two important implications. First, it seemed entirely clear that the course of Graves' disease was not related to the concentration of LATS in the plasma. Second, since the thyroid became overactive the case provides no support for the argument that occasional patients with high blood LATS concentrations fail to have thyrotoxicosis because of lymphocytic replacement in the gland by a Hashimoto-like process. CHOPRA and SOLOMON (1970) suggested that perhaps the thyrotoxicosis was due to an intrinsic abnormality of the thyroid.

A role for LATS in the origin of Graves' ophthalmopathy is not established (MCKENZIE and MCCULLAGH, 1968). KRISS et al. (1967) reported three patients in whom ophthalmopathy followed the appearance of circulating LATS, but this sequence was not found by BURKE in his group of patients (BURKE and SILVERSTEIN, 1969). Furthermore, there is frequently little relationship between symptoms of thyroid activity and ocular involvement.

Similarly, the reports of very high titers of circulating LATS in some patients with pretibial myxedema are counterbalanced by others with high titers and no pretibial myxedema, and by patients with pretibial myxedema and no circulating LATS.

One is left, then, without an explanation for the role of LATS in Graves' disease or for the curious phenomenon that this gamma globulin appears in the blood of the majority of patients, but rarely if ever except in association with thyroid disease. All that one can say at present is that detection of LATS in the blood of a patient suspected of having Graves' disease is strong evidence favoring this diagnosis. Its absence does not exclude the disease nor does its presence indicate the severity of the disease, past, present, or future. Certainly, more information is needed in order to clarify the role of this factor in Graves' disease and to account for its presence in some of these patients.

V. Prophylaxis against Endemic Goiter with Iodized Oil

Endemic goiter continues to be a major health problem in most areas of the world where the disease has been known for generations. Substantial understanding of its pathogenesis, simple and effective preventive measures, and a rising world economy have done little to reduce its prevalence on a world-wide scale.

If endemic goiter implied only an enlargement of the thyroid, it could be dismissed as a trivial affliction. In fact, severe endemic goiter is accompanied by a substantial incidence of cretinism and deaf-mutism. Furthermore, there is good reason for suspecting that in many cases, the mental retardation (apart from overt cretinism), neuromuscular dsiturbance, and short stature that usually accompany severe endemic goiter may be causally related. In addition, the disease may have an unfavorable effect on general health and educational capacity.

Salt iodization has not proved practical in certain isolated parts of the world. One of the various methods for supplying dietary iodine has been the administration of iodized oil to establish long-term body stores. In New Guinea a number of years ago, a program was begun of goiter prevention by the injection of iodized poppy seed oil containing 37% iodine. Information available from those surbeys (HENNESSY, 1964; BUTTFIELD et al., 1965) indicates that the prophylactic programs were effective in reducing the prevalence of goiter and that the procedure was practical and safe. The earlier results obtained in New Guinea did not provide enough information, however, regarding the effectiveness of this form of prophylaxis in reducing the incidence of cretinism and gave virtually no information regarding prevention of those disabilities epidemiologically associated with severe endemic goiter.

More recently, measurements of plasma TSH have been made before and several weeks to months after administration of single injections of iodized oil to goitrous natives in New Guinea (BUTTFIELD and HETZEL, 1969). Most of the subjects chosen had moderate to marked elevations of plasma TSH before the injection, and in almost all cases TSH returned to normal or near normal values afterward.

It is difficult to assess the effectiveness of any preventive program on the incidence of cretinism. This problem arises because of the long time-span of observation required, the relative rarity of the disease, the difficulty of making an early diagnosis because of progression of findings during the early months and years, and the high mortality of these persons in early life. BUTTFIELD and his colleagues (BUTTFIELD et al., 1970) attempted to trace the history of the mothers of cretins in New Guinea to verify the results of their iodized oil program. Although a few cretins were born of mothers who had been given iodized oil, in every case the injection had been late in pregnancy.

DELANGE et al. (1969) have exploited an interesting situation on Idjwi Island in the Congo. Severe endemic goiter and cretinism are found in the northern part of the island and little or none in the southern half, with few or no differences in social or dietary factors. Iodine intake is low and approximately uniform. Within a year after prophylactic injection of iodized oil, the prevalence of goiter had fallen to one-third its initial level, and in 80% of these cases the improvement was marked. The excretion rate of iodine was well maintained for a least a year, despite the small doses of iodized oil administered. No untoward effects were observed.

An extensive study of the effectiveness of iodized oil prophylaxis was begun in certain rural villages in highland Ecuador and Peru in 1966. Much information on the results of these programs has now accumulated (FIERRO-BENITEZ et al., 1969; PRETELL et al., 1969). It is clear that goiter incidence and magnitude is much reduced but not eliminated in adults who already had advanced cases, and it is virtually eliminated in the young. There has been little clear evidence of an effect on growth and development, but some evidence of improved intellectual performance in the young when compared to control groups. A few older persons with large nodular goiters have developed thyrotoxicosis, presumably because these nodules became independent of pituitary control and synthesized large amounts of hormone when they were supplied with abundant iodide from the injected iodized oil. In the study in Ecuador, six cretins have been identified in the control group of approximately 2500 persons, and none in the group of about 900 who received iodized oil (FIERRO-BENITEZ et al., 1970). It is interesting that in the Peruvian study, the usual rise of protein-bound iodine concentration failed to occur during pregnancy in the control group, but the usual rise did occur in women who had received iodized oil (PRETELL et al., 1969).

These findings indicate that injection of iodized poppy seed oil is an effective and safe method of preventing endemic goiter and its sequelae. It is effective for as long as 3 years after a single administration. Complications are rare and manageable. It has been recommended for use in situations where prophylaxis by salt iodization is impractical (KEVANY et al., 1969).

The effectiveness of a prophylactic program for endemic goiter does not exclude the possibility that other dietary, genetic, or environmental factors may modify the effect of iodide deficiency and the appearance of goiter, cretinism, and other associated defects. A number of observations suggest that other factors play a role. For example, there is a poor correlation between the severity of iodide deficiency in a population group and incidence of cretinism or deaf-mutism. GAITAN (GATIAN et al., 1969) in Colombia has detected a factor in well water that has an antithyroid effect, and EKPECHI (1967) in Nigeria has shown that rats fed cassava develop enlarged thyroids.

Differences in the clinical characteristics of cretinism in different endemias also suggest the presence of modifying factors in addition to the prevalent iodide deficiency. The cretins in most severe endemias have a striking resemblance to each other. The patient is usually not clinically hypothyroid or is only mildly hypothyroid, yet neurological deficits are striking and usually a large goiter is present. An exception is the endemic cretin of the Congo, so well described by DUMONT (DUMONT et al., 1963) who termed him "myxedematous". They differ from the more familiar type of cretin in several aspects: bone age is more strikingly delayed, they are clinically myxedematous or hypothyroid, stature is more retarded, and they appear not to be as impaired intellectually. It seems possible that local dietary factors or a combination of protein deficiency with a high intake of cyanogenic glycosides may be unusually damaging to fetal development and perinatal growth. These possibilities are under intensive investigation.

Nothing that has been discussed above should be taken as an indication that iodine prophylaxis for endemic goiter should not be instituted. There has been no significant residue of thyroid disease or its associated disorders in any country or any region where a prophylactic program has been seriously and carefully executed.

VI. Inborn Errors of the Thyroid

During the past 20 years, a number of disorders of the thyroid have been discovered that have been attributed with some credibility to inborn errors in the metabolism of the thyroid. The field has been reviewed from several points of view in the past few years (Dumont et al., 1963; Collaço, 1968; Hutchison, 1969; Stanbury, 1972). There are now approximately ten

Table 2. Inherited defects in the thyroid hormone system. Only the essential laboratory findings are given. Diagnosis requires additional information and usually detailed biochemical examination of a biopsy specimen

Disorder	Defect	Principal findings	
		clinical	laboratory
Transport defect	I^- transport	hypothyroid; goiter	No I^- transport into thyroid or salivary glands
"Peroxidase" defect	I^- organification	hypothyroid; goiter	SCN^- dumping of thyroid I^-
Pendred syndrome	I^- organification	hypothyroid; goiter; nerve deafness	Partial SCN^- dumping of thyroid I^-
Dehalogenase defect	MIT and DIT deiodination	eu- or hypothyroid; goiter	No deiodination of MIT or DIT; DI*T in urine after I* or DI*T
"Coupling" defect	T_3 and T_4 synthesis	eu- or hypothyroid; goiter	High I* uptake; low hormone production
TSH unresponsiveness	response to TSH	hypothyroid; no goiter	High blood TSH; low I* uptake
Thyroglobulin failure	thyroglobulin synthesis	hypothyroid; goiter	High I* uptake and turnover; no thyroglobulin
Hormone resistance	response to T_3 and T_4	eu- or hypothyroid; stipled epiphyses; deafness	High blood T_4; high I* uptake
Iodoprotein defect	?	eu- or hypothyroid; goiter	Iodoalbumin or iodoprealbumin in blood
Protease defect	thyroglobulin degradation	hypothyroid; goiter	Reduced protease activity in thyroid biopsy

species of thyroid disease that can be described in terms of more or less specific, identifiable biochemical disorders at a specific step in metabolism (Table 2). The complexity of thyroid hormone synthesis, storage, and secretion and the intricate chain of events from iodide uptake by the thyroid to the effect of the finished hormone in the peripheral cell clearly encompass a vast number of possibilities for metabolic error. Since survival without thyroid function is often possible and since compensation for metabolic error of the thyroid can be achieved through growth of the thyroid gland, one may be frequently confronted with a patient whose thyroid disease arises from an inherited disturbance of the thyroid system.

It is not possible to review these defects here, but only to point out some of the recent observations that have been made in the field. PAPADOPOULOS et al. (1970) have described a patient with a partial defect in iodide transport into the thytoid. Partial failure of this critical initial step in hormone biosynthesis impaired net thyroid function.

The normal thyroid recovers iodide from mono- and diiodotyrosine in the process of thyroglobulin degradation. When this deiodinating activity is absent or impaired, hormone synthesis is also impaired. LISSITZKY et al. (1968) described a patient with congenital goiter and an abnormal iodinated protein in the thyroid that they attributed to a possible defect in deiodinating activity. This interpretation has been challenged by THOMSON and MCGIRR (1969). Late appearance of the dehalogenase defect was observed in a 55-year-old male with goiter (JAFFIOL et al., 1969). HARDEN et al. (1967) have examined the role of plasma inorganic iodine on thyroid function in the dehalogenase deficiency syndrome, and patients with this disorder have been reported by NIALL et al. (1968) and MURRAY et al. (1965).

Among its many metabolic functions the thyroid must synthesize and store protein, especially thyroglobulin, and couple hormone precursors into finished hormone products. ALEXANDER and BURROW (1970) reported finding three cretins with goiter who appeared to be unable to synthesize thyroglobulin.

The thyroid gland of a patient with familial goiter and hypothyroidism, studied in the author's laboratory, also proved to have impaired thyroglobulin synthesis (RIDDICK et al., 1969). Using a variety of criteria, we found no evidence for the presence of thyroglobulin. From these studies, the fundamental defect responsible for the clinical state of this patient appeared to be the inability to synthesize this protein. The author knows of several other patients who apparently have the same defect but who, on histological staining of thyroid slices, showed no colloid staining in the thyroid follicles. This suggests the possibility that the colloid spaces were filled with unusual material that had solubility characteristics permitting its loss during the processes of tissue fixation or staining.

The thyroid proteins of a goitrous cretin with a defect in peroxidative covalent linking of iodide to tyrosyl residues have been studied by Mouriz et al. (1969). Sixteen percent of the soluble protein of the thyroid homogenate was identified immunologically as thyroglobulin, but this component comprised two fractions with sedimentation constants of 17 and 11.8. Much of the soluble protein was hemoglobin. The rest was of low molecular weight.

The iodoproteins of the thyroid have been studied by Medeiros-Neto et al. (1968) in two patients with the Pendred syndrome (familial goiter and deafness), but there may be some uncertainty whether these were true instances of the Pendred syndrome or deaf persons who also had endemic goiter. The thyroidal iodoproteins in congenital goiter of calves from South Africa have been investigated by Van Zyl and his colleagues (Robbins et al., 1966).

Furth et al. (1970) have described a cretinous woman with goiter who had a large amount of iodoprealbumin in the blood but thyroglobulin in the thyroid. They attributed the disorder to absence of proteolytic enzymes in the thyroid for degrading thyroglobulin, but other explanations are possible. Familial goiter with unresponsiveness to thyroid hormone has been reported by Refetoff et al. (1967). An unusual family syndrome of goiter with intrathyroidal calcification has been described by Murray et al. (1966). Just what metabolic abnormality of thyroid function these last patients have remains to be determined.

An 8-year-old cretin with TSH unresponsiveness and without goiter has been studied in the author's laboratory. At biopsy the thyroid proved to be normal in size (Stanbury et al., 1968). The biopsy specimen showed a normal monoiodotyrosine-diiodotyrosine (MIT/DIT) ratio and thyroxine. The gland failed to respond to TSH or to produce thyroglobulin in both *in vivo* and *in vitro* studies. The inference was drawn that the defect in this patient was a failure of the thyroid to respond to TSH in its two characteristic modalities, namely, cell division and thyroglobulin synthesis.

Two findings of interest have been described in a patient from a family with the Pendred syndrome (Milutinovic et al., 1969). Approximately 25% of the labeled iodine was discharged from the gland when perchlorate was given 3 weeks after the labeling dose of radioiodine. This effect indicated the possibility that the gland had a large persistent pool of inorganic iodine. The second finding was an unusually high MIT/DIT ratio in a gland in which the content of iodine per gram of tissue was within normal limits 3 weeks after the labeling iodine had been given and when equilibrium should have been approached. Evidence was also obtained for a large spillage of nonhormonal iodine from the gland. Thyroglobulin was present, however. This unusual sensitivity to perchlorate may prove to be a new and biologically significant phenomenon in the pathogenesis of the Pendred syndrome.

References

ADAMS, D. D.: The presence of an abnormal thyroid-stimulating hormone in the serum of some thyrotoxic patients. J. clin. Endocr. **18**, 699–712 (1958).

ADAMS, D. D.: A comparison of the rates at which thyrotrophin and the human abnormal thyroid stimulator disappear from the circulating blood of the rat. Endocrinology **66**, 658–664 (1960).

ADAMS, D. D., COUCHMAN, K., KILPATRICK, J. A.: Lack of response to TSH injections in euthyroid patients with high LATS levels. J. clin. Endocr. **29**, 1502–1503 (1969).

ADAMS, D. D., KENNEDY, T. H., PURVES, H. D., SIRETT, N. E.: Failure of TSH antisera to neutralize long-acting thyroid stimulator. Endocrinology **70**, 801–805 (1962).

AHN, C. S., ATHANS, J. C., ROSENBERG, I. N.: Stimulation of thyroid hormone secretion by dibutyryl cyclic-AMP. Endocrinology **85**, 224–230 (1969).

AHN, C. S., ROSENBERG, I. N.: Proteolysis in thyroid slices: Effects of TSH, dibutyryl cyclic 3',5'-AMP and prostaglandin E_1. Endocrinology **86**, 870–873 (1970a).

AHN, C. S., ROSENBERG, I. N.: Iodine metabolism in thyroid slices: Effects of TSH dibutyryl cyclic 3',5'-AMP, NaF and prostaglandin E_1. Endocrinology **86**, 396–405 (1970b).

ALEXANDER, N. M., BURROW, G. N.: Thyroxine biosynthesis in human goitrous cretinism. J. clin. Endocr. **30**, 308–315 (1970).

AVERILL, R. L. W.: Responses to thyrotropin-releasing factor (TRF) by intrapituitary infusion of hypothalamic extracts. Endocrinology **84**, 514–519 (1969a).

AVERILL, R. L. W.: Depression of thyrotropin releasing factor induction of thyrotropin release by thyroxine in small doses. Endocrinology **85**, 67–71 (1969b).

BAUGH, C. M., KRUMDIECK, C. L., HERSHMAN, J. M., PITTMAN, J. A., JR.: Synthesis and biological activity of thyrotropin-releasing hormone. Endocrinology **87**, 1015 (1970).

BEALL, G. N., DANIEL, P. M., PRATT, O. E., SOLOMON, D. H.: Effects of immunization of baboons with human thyroid tissue. J. clin. Endocr. **29**, 1460–1469 (1969b).

BEALL, G., DONIACH, D., ROITT, I., EL KABIR, D.: Inhibition of the long-acting thyroid stimulator (LATS) by soluble thyroid fractions. J. Lab. clin. Med. **73**, 988–999 (1969a).

BOWERS, C. Y., LEE, K. L., SCHALLY, A. V.: Effect of actinomycin D on hormones that control the release of thyrotropin from the anterior pituitary glands of mice. Endocrinology **82**, 303–310 (1968b).

BOWERS, C. Y., LEE, K. L., SCHALLY, A. V.: A study on the interaction of the thyrotropin-releasing factor and L-triiodothyronine: Effects of puromycin and cycloheximide. Endocrinology **82**, 75–82 (1968c).

BOWERS, C. Y., SCHALLY, A. V., ENZMANN, F., BOLER, J., FOLKERS, K.: Porcine thyrotropin releasing hormone is (pyro)glu-his-pro(NH_2). Endocrinology **86**, 1143–1153 (1970).

BOWERS, C. Y., SCHALLY, A. V., HAWLEY, W. D., GUAL, C., PARLOW, A.: Effect of thyrotropin-releasing factor in man. J. clin. Endocr. **28**, 978–982 (1968a).

BURGUS, R., DUNN, T. F., DESIDERIO, D. M., WARD, D. N., VALE, W., GUILLEMIN, R., FELIX, A. M., GILLESSEN, D., STUDER, R. O.: Biological activity of synthetic polypeptide derivatives related to the structure of hypothalamic TRF. Endocrinology **86**, 573–582 (1970).

BURKE, G.: The long-acting thyroid stimulator of Graves' disease. Amer. J. Med. **45**, 435–450 (1968a).

BURKE, G.: On the competitive interaction of long-acting thyroid stimulator and thyrotropin *in vivo*. J. clin. Endocr. **28**, 286–293 (1968b).

BURKE, G.: Effects of adrenergic blocking agents on basal and stimulated thyroid function. Metabolism **18**, 961–967 (1969a).

BURKE, G.: Failure of theophylline to potentiate stimulated thyroidal glucose oxidation and phospholipogenesis. Endocrinology **84**, 1055–1062 (1969b).

BURKE, G.: Comparative effects of thyrotropin and long-acting thyroid stimulator on thyroidal intermediary metabolism: Relationship to pyridine nucleotide levels. Metabolism **18**, 132–140 (1969c).

BURKE, G.: The cell membrane: A common site of action of thyrotropin (TSH) and long-acting thyroid stimulator (LATS). Metabolism **18**, 720–729 (1969d).

Burke, G.: Comparison of thyrotropin and sodium fluoride effects on thyroid adenyl cyclase. Endocrinology **86**, 346–352 (1970a).

Burke, G.: On the role of adenyl cyclase activation and endocytosis in thyroid slice metabolism. Endocrinology **86**, 353–359 (1970b).

Burke, G., Silverstein, G. E.: Hypothyroidism after treatment with sodium iodide I-131. J. Amer. med. Ass. **210**, 1051–1058 (1969).

Buttfield, I. H., Black, M. L., Hoffman, M. J., Mason, E. K., Hetzel, B. S.: Correction of iodine deficiency in New Guinea natives by iodised oil injection. Lancet **1965 II**, 767–769.

Buttfield, I. H., Hetzel, B. S.: Endemic goiter in New Guinea and the prophylactic program with iodinated poppyseed oil. In: Endemic goiter (J. B. Stanbury, ed.), chapter 11, p. 132. Washington, D.C.: WHO 1969.

Buttfield, I. H., Pharoah, P. O. D., Hetzel, B. S.: Evidence of prevention of neurological defect in New Guinea children by iodised oil injection of mothers prior to pregnancy. Presented at the 6th International Thyroid Conference, Vienna, June 1970. Abstract No 3, p. 15.

Carneiro, L., Dorrington, K. J., Munro, D. S.: Recovery of the long-acting thyroid stimulator from serum of patients with thyrotoxicosis by concentration of immunoglobulin G. Clin. Sci. **31**, 215–221 (1966a).

Carneiro, L., Dorrington, K. J., Munro, D. S.: Relation between long-acting thyroid stimulator and thyroid function in thyrotoxicosis. Lancet **1966 II**, 878–880.

Cavalieri, R. R., Searle, G. L., Rosenberg, L. L.: Studies of thyroglobulin of hypophysectomized rats. II. Nature of thyroidal iodoprotein after hypophysectomy of rats previously depleted of colloid by propylthiouracil. Endocrinology **86**, 10–17 (1970).

Chopra, I. J., Solomon, D. H.: Graves' disease with delayed hyperthyroidism. Ann. intern. Med. **73**, 985 (1970).

Chopra, I. J., Solomon, D. H., Limberg, N. P.: Specific and nonspecific responses in the bioassay of long-acting thyroid stimulator (LATS). J. clin. Endocr. **31**, 382–390 (1970).

Clayton, J. A., Masuoka, D. T.: TSH-induced mobilization of serotonin from perivascular mast cells in the rat thyroid. Endocrinology **83**, 263–271 (1968).

Clayton, J. A., Szego, C. M.: Depletion of rat thyroid serotonin accompanied by increased blood flow as an acute response to thyroid-stimulating hormone. Endocrinology **80**, 689–698 (1967).

Collaço, F. M.: Bocio por dishormonogenesis. Rev. clín. esp. **108**, 490 (1968).

Conway, L. W., Schalch, D. S., Utiger, R. D., Reichlin, S.: Hormones in human pituitary sinusoid blood: Concentration of LH, GH and TSH. J. clin. Endocr. **29**, 446–456 (1969).

Delange, F., Thilly, C., Pourbaix, P., Ermans, A. M.: Treatment of Idjwi Island endemic goiter by iodized oil. In: Endemic goiter (J. B. Stanbury, ed.), chapter 10, p. 118. Washington, D.C.: WHO 1969.

Dumont, J. E.: Le mécanisme d'action de l'hormone thyréotrope. Bull. Soc. Chim. biol. (Paris) **50**, 2401–2426 (1968).

Dumont, J. E., Ermans, A. M., Bastenie, P. A.: Thyroidal function in a goiter endemic. IV. Hypothyroidism and endemic cretinism. J. clin. Endocr. **23**, 325–335 (1963).

Dumont, J. F., Neve, P., Otten, J.: Recent advances in the knowledge of the control of thyroid growth and function. In: Endemic goiter (J. B. Stanbury, ed.), chapter 2, p. 14. Washington, D.C.: WHO 1969.

Ekpechi, O. L.: Pathogenesis of endemic goiter in Eastern Nigeria. Brit. J. Nutr. **21**, 537–545 (1967).

Ensor, J. M., Munro, D. S.: A comparison of the in-vitro actions of thyroid-stimulating hormone and cyclic 3′,5′-adenosine monophosphate on the mouse thyroid gland. J. Endocr. **43**, 477–485 (1969).

Falconer, I. R.: The effect of vasopressin on hormone secretion and blood flow from the thyroid vein in sheep with exteriorized thyroids. J. Physiol. (Lond.) **199**, 427–441 (1968).

Fierro-Benitez, R., Ramirez, I., Estrella, E., Jaramillo, C., Diaz, C., Urretsa, J.: Iodized oil in the prevention of endemic goiter and associated defects in the Andean region of Ecuador. In: Endemic goiter (J. B. Stanbury, ed.), chapter 26, p. 306. Washington, D.C.: WHO 1969.

FIERRO-BENITEZ, R., RAMIREZ, I., ESTRELLA, E., QUERIDO, A., STANBURY, J. B.: The effect of goiter prophylaxis with iodized oil on the prevention of endemic cretinism. Presented at the 6th International Thyroid Conference, Vienna, June 1970. Abstract No 4, p. 16.

FLEISCHER, N., BURGUS, R., VALE, W., DUNN, T., GUILLEMIN, R.: Preliminary observations on the effect of synthetic thyrotropin releasing factor on plasma thyrotropin levels in man. J. clin. Endocr. **31**, 109–112 (1970).

FOLKERS, K., ENZMANN, F., BOLER, J., BOWERS, C. Y., SCHALLY, A. V.: Discovery of modification of the synthetic tripeptide-sequence of the thyrotropin releasing hormone having activity. Biochem. biophys. Res. Commun. **37**, 123–126 (1969).

FUJITA, H., SUEMASA, H.: Cytological effects of TSH on the thyroid of hypophysectomized rats with and without previous administration of Actinomycin D. An electron microscope study. Arch. histol. jap. **30**, 45–59 (1968).

FURTH, E. D., AGRAWAL, R. B., PROPP, R. P.: Secretion of iodoalbumin and iodoprealbumin by a congenital goiter containing thyroglobulin and the iodoalbumins. J. clin. Endocr. **31**, 60–69 (1970).

GAITAN, E., WAHNER, H. W., CUELLA, C., CORREA, P., JUBIZ, W., GAITAN, J. E.: Endemic goiter in the Cauca Valley: II. Studies of thyroid pathophysiology. J. clin. Endocr. **29**, 675–683 (1969).

GILMAN, A. G., RALL, T. W.: Factors influencing adenosine 3',5'-phosphate accumulation in bovine thyroid slices. J. biol. Chem. **243**, 5867–5871 (1968)a.

GILMAN, A. G., RALL, T. W.: The role of adenosine 3',5'-phosphate in mediating effects of thyroid-stimulating hormone on carbohydrate metabolism of bovine thyroid slices. J. biol. Chem. **243**, 5872–5881 (1968b).

GREER, M. A., GRIMM, Y., STUDER, H.: Qualitative changes in the secretion of thyroid hormones induced by iodine deficiency. Endocrinology **83**, 1193–1198 (1968).

GREER, M. A., ROCKIE, C.: Effect of thyrotropin and the iodine content of the thyroid on the triiodothyronine: thyroxine ratio of newly synthesized iodothyronines. Endocrinology **85**, 244–250 (1969).

HARDEN, R. McG., ALEXANDER, W. D., PAPADOPOULOS, S., HARRISON, M. T., MACFARLANE, S.: The influence of the plasma inorganic iodine concentration on thyroid function in dehalogenase deficiency. Acta endocr. (Kbh.) **55**, 361–368 (1967).

HART, I. R., McKENZIE, J. M.: Comparison of the effects of thyrotropin and the long-acting thyroid stimulator on guinea pig adipose tissue. Endocrinology **88**, 26 (1971).

HENNESSY, W. B.: Goitre prophylaxis in New Guinea with intramuscular injections of iodized oil. Med. J. Aust. **1**, 505–512 (1964).

HODGE, J. V., MELMON, K. L., SJOERDSMA, A.: Hormone-induced modifications of free tyrosine in the rat thyroid gland. J. Physiol. (Lond.) **203**, 1–12 (1969).

HULLIN, R. P., JOHNSON, A. W.: Effect of lithium salts on uptake of I^{125} by rat thyroid gland. Life Sci. **9**, 9–20 (1970).

HUTCHISON, J. H.: Familial goitrous hypothyroidism. In: Endocrine and genetic diseases of childhood (L. I. GARDNER, ed.), p. 253. Philadelphia: Saunders 1969.

IMBENOTTE, J., NATAF, B., HAREL, J.: Action de la TSH sur la biosynthèse du RNA dans des thyroïdes de foetus de rat en culture organotypique. Bull. Soc. Chim. biol. (Paris) **51**, 428–432 (1969).

JACQUEMIN, C., HAYE, B.: Controle de l'activité biologique de la thyreostimuline par la thyroxine. C. R. Soc. Biol. (Paris) **162**, 1064–1069 (1968).

JAFFIOL, C., KHALIL, R., PASTORELLO, R., BALDET, L., MIROUZE, J.: Goitre sporadique tardif de l'adulte, avec déficit enzymatique en deshalogénase. Étude clinique, biochimique, isotopique. Rev. franç. Endocr. clin. **10**, 67–71 (1969).

KANEKO, T., ZOR, U., FIELD, J. B.: Thyroid-stimulating hormone and prostaglandin E_1 stimulation of cyclic 3',5'-adenosine monophosphate in thyroid slices. Science **163**, 1062–1063 (1969).

KERKOF, P. R., TATA, J. R.: The subcellular distribution of ^{32}P-labelled phospholipids, ^{32}P-labelled ribonucleic acid and ^{125}I-labelled iodoprotein in pig thyroid slices. Biochem. J. **112**, 729–739 (1969).

KEVANY, J., FIERRO-BENITEZ, R., PRETELL, E. A., STANBURY, J. B.: Prophylaxis and treatment of endemic goiter with iodized oil in rural Ecuador and Peru. Amer. J. clin. Nutr. **22**, 1597–1607 (1969).

Kohler, P. O., Bridson, W. E., Rayford, P. L., Kohler, S. E.: Hormone production by human pituitary adenomas in culture. Metabolism **18**, 782–788 (1969).

Konno, N., Murthy, P. V. N., McKenzie, J. M.: Stimulation of proteolysis in the mouse thyroid gland by thyrotropin and the long-acting thyroid stimulator: Comparison of intact lobe and homogenate. Endocrinology **87**, 1062 (1970).

Krass, M. E., LaBella, F. S., Mailhot, R.: Bovine thyrotropin: molecular weight of the hormone derived from soluble and granular pituitary fractions, and preparation of an active fragment by enzyme hydrolysis. Endocrinology **84**, 1257–1261 (1969).

Krass, M. E., LaBella, F. S., Vivian, S. R.: Thyrotropin release *in vitro:* The role of metabolism in the secretory response to vasopressin, oxytocin and epinephrine. Endocrinology **82**, 1183–1189 (1968).

Kriss, J. P.: Inactivation of long-acting thyroid stimulator (LATS) by anti-kappa and anti-lambda antisera. J. clin. Endocr. **28**, 1440–1444 (1968).

Kriss, J. P., Pleshakov, V., Rosenblum, A. L., Holderness, M., Sharp, G., Utiger, R.: Studies on the pathogenesis of the ophthalmopathy of Graves' disease. J. clin. Endocr. **27**, 582–593 (1967).

Lee, K. L., Bowers, C. Y., Miller, O. N.: Some studies of the effect of thyroid hormone on the RNA synthesis of the anterior pituitary gland of rats. Endocrinology **83**, 763–768 (1968).

Levey, G. S., Roth, J., Pastan, I.: Effect of propranolol and phentolamine on canine and bovine responses to TSH. Endocrinology **84**, 1009–1015 (1969).

Lindsay, R. J., Cash, A. G., Hill, J. B.: TSH stimulation of orotic acid conversion to pyrimidine nucleotides and RNA in bovine thyroid. Endocrinology **84**, 534–543 (1969).

Lissitzky, S., Bismuth, J., Codaccioni, J.-L., Cartouzou, G.: Congenital goiter with iodoalbumin replacing thyroglobulin and defect of deiodination of iodotyrosines. Serum origin of the thyroid iodoalbumin. J. clin. Endocr. **28**, 1797–1806 (1968).

Lissitzky, S., Manté, S., Attali, J.-C., Cartouzou, G.: Action of 3′,5′-cyclic adenosine-monophosphate on the protein synthesizing capacity of thyroid polyribosomes *in vitro*. Biochem. biophys. Res. Commun. **35**, 437–443 (1969).

Loeb, L., Kaplan, E. E.: Studies on compensatory hypertrophy of the thyroid gland: effect of feeding anterior lobe of pituitary gland on the hypertrophy of thyroid gland in the guinea pig. J. med. Res. **44**, 557–578 (1924).

Maayan, M. L.: Effect of dinitrophenol on thyroid responses to thyrotropin. Endocrinology **83**, 938–944 (1968).

Maayan, M. L., Shapiro, R. J., Ingbar, S. H.: Metabolic functions of thyroid cell "ghosts". Presented at the 6th International Thyroid Conference, Vienna, June 1970. Abstract No 77, p. 89.

Macchia, V., Meldolesi, M. F., Maselli, P.: Effect of cyclic 3′,5′-AMP on glucose metabolism in thyroid homogenates. Endocrinology **85**, 895–898 (1969).

Macchia, V., Tamburrini, O., Pastan, I.: Role of lecithin the in mechanism of TSH action. Endocrinology **86**, 787–792 (1970).

Martin, J. B., Boshans, R., Reichlin, S.: Feedback regulation of TSH secretion in rats with hypothalamic lesions. Endocrinology **87**, 1032 (1970).

McKenzie, J. M.: Studies on the thyroid activator of hyperthyroidism. J. clin. Endocr. **21**, 635–647 (1961).

McKenzie, J. M.: Neonatal Graves' disease. J. clin. Endocr. **24**, 660–668 (1964).

McKenzie, J. M.: Humoral factors in the pathogenesis of Graves' disease. Physiol. Rev. **48**, 252–310 (1968).

McKenzie, J. M., Adiga, P. R., Murthy, P. V. N.: Effects of actinomycin D, cycloheximide and puromycin on thyroid stimulation. Endocrinology **83**, 1132–1139 (1968).

McKenzie, J. M., Gordon, J.: The origin of the long-acting thyroid stimulator. In: Current topics in thyroid research (C. Cassano and M. Andreoli, eds.). Proceedings of the 5th International Thyroid Conference, pp. 445–454. New York: Academic Press, Inc. 1965.

McKenzie, J. M., McCullagh, E. P.: Observations against a caudal relationship between the long-acting thyroid stimulator and ophthalmopathy in Graves' disease. J. clin. Endocr. **28**, 1177–1182 (1968).

Medeiros-Neto, G. A., Nicolau, W., Kieffer, J., Cintra, A. B. U.: Thyroidal iodoproteins in Pendred's syndrome. J. clin. Endocr. **28**, 1205 (1968).

MESSIER, B.: Effect of exogenous thyroxine on ^{3}H-thymidine uptake in mouse pituitary gland. Acta endocr. (Kbh.) **61**, 133–136 (1969).

MILUTINOVIC, P. S., STANBURY, J. B., WICKEN, J. V., JONES, E. W.: Thyroid function in a family with the Pendred syndrome. J. clin. Endocr. **29**, 962–969 (1969).

MITTLER, J. C., REDDING, T. W., SCHALLY, A. V.: Stimulation of thyrotropin (TSH) secretion by TSH-releasing factor (TRF) in organ cultures of anterior pituitary. Proc. Soc. exp. Biol. (N.Y.) **130**, 406–409 (1969).

MIYAI, K., FUKUCHI, M., KUMAHARA, Y., ABE, H.: LATS production by lymphocyte culture in patients with Graves' disease. J. clin. Endocr. **27**, 855–860 (1967).

MOURIZ, J., RIESCO, G., USOBIAGA, P.: Thyroid proteins in a goitrous cretin with iodide organification defect. J. clin. Endocr. **29**, 942–947 (1969).

MURRAY, I. P. C., THOMSON, J. A., MCGIRR, E. M., MACDONALD, E. M., KENNEDY, J. S., MCLENNAN, I.: Unusual familial goiter associated with intrathyroidal calcification. J. clin. Endocr. **26**, 1039–1049 (1966).

MURRAY, P., THOMSON, J. A., MCGIRR, E. M., WALLACE, T. J.: Absent and defective iodotyrosine deiodination in a family some whose members are goitrous cretins. Lancet **1965 I**, 183–185.

NIALL, H. D., WELLBY, M. L., HETZEL, B. S., HUDSON, B., CHENOWETH, R. A.: Biochemical and clinical studies in familial goitre caused by an iodotyrosine deiodinase defect. Aust. Ann. Med. **17**, 89–95 (1968).

OCHI, Y., DEGROOT, L. J.: Stimulation of RNA and phospholipid formation by long acting thyroid stimulator and by thyroid-stimulating hormone. Biochim. biophys. Acta (Amst.) **170**, 198–201 (1968).

OCHI, Y., DEGROOT, L. J.: Effect of adenine nucleotides on thyroid hormone release in vivo and in vitro. Metabolism **18**, 331–338 (1969a).

OCHI, Y., DEGROOT, L. J.: Stimulation of thyroid hyperplasia and protein synthesis by LATS. Endocrinology **85**, 344–347 (1969b).

ONAYA, T., SOLOMON, D. H.: Effects of chlorpromazine and propranolol on *in vitro* thyroid activation by thyrotropin, long-acting thyroid stimulator and dibutyryl cyclic-AMP. Endocrinology **85**, 1010–1017 (1969).

ONAYA, T., SOLOMON, D. H.: Stimulation by prostaglandin E_1 of endocytosis and glucose oxidation in canine thyroid slices. Endocrinology **86**, 423–425 (1970).

PAPADOPOULOS, S. N., VAGENAKIS, A. G., MOSCHOS, A., KOUTRAS, D. A., MATSANIOTIS, N., MALAMOS, B., BISMUTH, J., BECHET, M. M., LISSITZKY, S.: A case of a partial defect of the iodide trapping mechanism. J. clin. Endocr. **30**, 302–307 (1970).

PASTAN, I., KATZEN, R.: Activation of adenyl cyclase in thyroid homogenates by thyroid-stimulating hormone. Biochem. biophys. Res. Commun. **29**, 792–798 (1967).

PASTAN, I., ROTH, J., MACCHIA, V.: Binding of hormone to tissue: the first step in polypeptide hormone action. Proc. nat. Acad. Sci. (Wash.) **56**, 1802–1809 (1966).

PASTAN, I., WOLLMAN, S. H.: Colloid droplet formation in dog thyroid in vitro. J. Cell Biol. **35**, 262–266 (1967).

PINCHERA, A., LIBERTI, P., BADALAMENTI, G.: Attività tireostimolante ad azione prolungata nel siero di conigli immunizzati con tiroide umana. Folia endocr. (Roma) **18**, 522–534 (1965).

PRETELL, E. A., MONCLOA, F., SALINAS, R., KAWANO, A., GUERRA-GARCIA, R., GUTIERREZ, L., BETETA, L., PRETELL, J., WAN, M.: Prophylaxis and treatment of endemic goiter in Peru with iodized oil. J. clin. Endocr. **29**, 1586–1595 (1969).

READ, D. G., HERSHMAN, J. M., PITTMAN, J. A.: Effect of vasopressin infusions on thyroidal radioiodine uptake and serum TSH concentration. J. clin. Endocr. **29**, 1496–1498 (1969).

REDDING, T. W., SCHALLY, A. V.: Studies on the thyrotropin-releasing hormone (TRH) activity in peripheral blood. Proc. Soc. exp. Biol. (N.Y.) **131**, 420–425 (1969).

REFETOFF, S., DEWIND, L. T., DEGROOT, L. J.: Familial syndrome combining deaf-mutism, stippled epiphyses, goiter and abnormally high PBI: possible target organ refractoriness to thyroid hormone. J. clin. Endocr. **27**, 279–294 (1967).

RIDDICK, F. A., JR., DESAI, K. B., STANBURY, J. B., MURISON, P. J.: Familial goiter with diminished synthesis of thyroglobulin. Z. ges. exp. Med. **150**, 203–212 (1969).

ROBBINS, J., VAN ZYL, A., VAN DER WALT, K.: Abnormal thyroglobulin in congenital goiter of cattle. Endocrinology **78**, 1213–1223 (1966).

RODBELL, M.: Metabolism of isolated fat cells. V. Preparation of "ghosts" and their properties; adenyl cyclase and other enzymes. J. biol. Chem. **242**, 5744–5750 (1967).

RODESCH, F., NEVE, P., WILLEMS, C., DUMONT, J. E.: Stimulation of thyroid metabolism by thyrotropin, cyclic 3':5'-AMP, dibutyryl cyclic 3':5'-AMP and prostaglandin E_1. Europ. J. Biochem. **8**, 26–32 (1969).

ROGOWITSCH, N.: Die Veränderungen der Hypophyse nach Entfernung der Schilddrüse. Beitr. path. Anat. **4**, 453 (1888).

ROSENBERG, L. L., CAVALIERI, R. R.: Studies of thyroglobulin of hypophysectomized rats. I. Sensitivity to the disaggregating influences of alkali and low ionic strength. Endocrinology **84**, 1322–1329 (1969).

SCHALLY, A. V., REDDING, T. W., BOWERS, C. Y., BARRETT, J. F.: Isolation and properties of porcine thyrotropin-releasing hormone. J. biol. Chem. **244**, 4077–4088 (1969).

SCHELL-FREDERICK, E., DUMONT, J. E.: Mechanism of action of thyrotropin. In: Biochemical actions of hormones (G. LITWACK, ed.), vol. 1, chapter 10, p. 415. New York: Academic Press, Inc. 1970.

SCHIMMER, B. P., UEDA, K., SATO, G. H.: Site of action of adrenocorticotropic hormone (ACTH) in adrenal cell cultures. Biochem. biophys. Res. Commun. **32**, 806–810 (1968).

SCHNEIDER, P. B.: Effects of thyrotropin on thyroidal phospholipid and adenosine 5'-triphosphate metabolism. J. biol. Chem. **244**, 4490–4493 (1969).

SCOTT, T. W., MILLS, S. C., FREINKEL, N.: The mechanism of thyrotropin action in relation to lipid metabolism in thyroid tissue. Biochem. J. **109**, 325–332 (1968).

SHISHIBA, Y., SOLOMON, D. H., DAVIDSON, W. D.: Comparison of the effect of thyrotropin and long-acting thyroid stimulator on glucose oxidation and endocytosis in canine thyroid slices. Endocrinology **86**, 183–190 (1970).

SINGH, S. P., MCKENZIE, J. M.: Effects of thyrotropin on ^{35}S-sulfate uptake by mouse harderian gland. Endocrinology **85**, 952–955 (1969).

SMITH, B. R.: The interaction of the long-acting thyroid stimulator (LATS) with thyroid tissue *in vitro*. J. Endocr. **46**, 45–54 (1970).

SMITH, B. R., MUNRO, D. S., DORRINGTON, K. J.: The distribution of the long-acting thyroid stimulator among γG-immunoglobulins. Biochim. biophys. Acta (Amst.) **188**, 89–100 (1969).

SOLOMON, D. H., BEALL, G. N.: Thyroid-stimulating activity in the serum of immunized rabbits. II. Nature of the thyroid-stimulating material. J. clin. Endocr. **28**, 1496–1502 (1968).

SOLOMON, D. H., BEALL, G. N.: Effect of thyroxine on the thyroid-stimulating activity in the serum of rabbits immunized with thyroid tissue. Endocrinology **86**, 191–195 (1970).

SPEIGHT, J. W., BABA, W. I., WILSON, G. M.: The effect of propylthiouracil and thyroid-stimulating hormone on the survival of rat thyroid cells *in vivo* and *in vitro*. J. Endocr. **41**, 577–591 (1968).

STANBURY, J. B., ROCMANS, P., BUHLER, U. K., OCHI, Y.: Congenital hypothyroidism with impaired thyroid response to thyrotropin. New Engl. J. Med. **279**, 1132–1136 (1968).

STANBURY, J. B., WICKEN, J. V., LAFFERTY, M. A.: Preparation and properties of thyroid cell membranes. J. Membrane Biol. **1**, 459–467 (1969).

STANBURY, J. B., WICKEN, J. V., PARDO, N.: Unpublished observations.

STANBURY, J. B., WYNGAARDEN, J. B., FREDRICKSON, D. S. (eds.): The Metabolic Basis of Inherited Disease. 3rd edition. McGraw-Hill, New York, 1972.

STRAZNICKY, K., MESS, B.: Effect of TSH on the rate of DNA and protein synthesis in embryonic chicken thyroid cells, investigated with ^{3}H-thymidine and ^{3}H-methionine autoradiography. Acta biol. Acad. Sci. hung. **18** (2), 221–229 (1967).

SUEMATSU, R., MATSUDA, K., SHIZUME, K., NAKAO, K.: Thyroid response to acute reduction of circulating thyroid hormone level. Endocrinology **84**, 1161–1165 (1969).

SUNSHINE, P., KUSUMOTO, H., KRISS, I. P.: Survival time of circulating long-acting thyroid stimulator in neonatal thyrotoxicosis: Implications for diagnosis and therapy of the disorder. Pediatrics **36**, 869–876 (1965).

TAUNTON, O. D., ROTH, J., PASTAN, I.: The first step in ACTH action: Binding to tissue. J. clin. Invest. **46**, 1122 (1967).

THOMSON, J. A., McGIRR, E. M.: Defective deiodinase activity and abnormal thyroidal iodoproteins. J. clin. Endocr. **29**, 1259–1260 (1969).

TONG, W.: TSH stimulation of ^{14}C-amino acid incorporation into protein by isolated bovine thyroid cells. Endocrinology **80**, 1101–1110 (1967).

TONOUE, T., TONG, W., STOLC, V.: TSH and dibutyryl-cyclic-AMP stimulation of hormone release from rat thyroid glands *in vitro*. Endocrinology **86**, 271–277 (1970).

VALE, W., BURGUS, R., DUNN, T. F., GUILLEMIN, R.: Release of TSH by oral administration of synthetic peptide derivatives with TRF activity. J. clin. Endocr. **30**, 148–150 (1970).

WILBER, J. F., PEAKE, G. T., UTIGER, R. D.: Thyrotropin release in vitro: Stimulation by cyclic 3′,5′-adenosine monophosphate. Endocrinology **84**, 758–760 (1969).

WILBER, J. F., PORTER, J. C.: Thyrotropin and growth hormone releasing activity in hypophysial portal blood. Endocrinology **87**, 807–811 (1970).

WILBER, J. F., UTIGER, R. D.: Thyrotropin incorporation of ^{14}C-glucosamine by the isolated rat adenohypophysis. Endocrinology **84**, 1316–1321 (1969a).

WILBER, J. F., UTIGER, R. D.: The effect of glucocorticoids on thyrotropin secretion. J. clin. Invest. **48**, 2096–2103 (1969b).

WILLEMS, C., RODESCH, F., NEVE, P., DUMONT, J. E.: Mechanism of action of thyrotropin through the cyclic 3′,5′-AMP system. Arch. int. Physiol. **77** (1), 179–180 (1969).

WILLIAMS, J. A.: Effects of TSH on thyroid membrane properties. Endocrinology **86**, 1154–1158 (1970).

WILLIAMS, J. A., WOLFF, J.: Thyroid secretion in vitro: Multiple actions of agents affecting secretion. Endocrinology **88**, 206 (1971).

WILSON, B., RAGHUPATHY, R., TONOUE, T., TONG, W.: TSH-like actions of dibutyryl-cAMP on isolated bovine thyroid cells. Endocrinology **83**, 877–884 (1968).

WOLFF, J.: Enzymatic properties of thyroid membranes. Presented at the 6th International Thyroid Conference, Vienna, June 1970, Abstract No 75, p. 81.

WOLFF, J., BERENS, S. C., JONES, A. B.: Inhibition of thyrotropin-stimulated adenyl cyclase activity of beef thyroid membranes by low concentration of lithium ion. Biochem. biophys. Res. Commun. **39**, 77–82 (1970).

WONG, E. T., LITMAN, G. W.: Interaction of purified long-acting thyroid stimulator (LATS) and thyroid microsomes *in vitro*. J. clin. Endocr. **29**, 72–78 (1969).

YAMASHITA, K., FIELD, J. B.: Preparation of thyroid plasma membranes containing a TSH-responsive adenyl cyclase. Biochem. biophys. Res. Commun. **40**, 171–178 (1970).

ZAKARIJA, M., BASTOMSKY, C. H., McKENZIE, J. M.: Effect of thyrotropin on thyroid pyridine nucleotides *in vivo*. Endocrinology **84**, 1310–1315 (1969).

ZOR, U., KANEKO, T., LOWE, I. P., BLOOM, G., FIELD, J. B.: Effect of thyroid-stimulating hormone and prostaglandins on thyroid adenyl cyclase activation and cyclic adenosine 3′,5′-monophosphate. J. biol. Chem. **244**, 5189–5195 (1969).

ZOR, U., LOWE, I. P., BLOOM, G., FIELD, J. B.: The role of calcium (Ca^{++}) in TSH and dibutyryl 3′5′ cyclic AMP stimulation of thyroid glucose oxidation and phospholipid synthesis. Biochem. biophys. Res. Commun. **33**, 649–658 (1968).

Psychophysical Basis of Coincidence Mechanisms in the Human Visual System

MAARTEN A. BOUMAN and JAN J. KOENDERINK*

With 22 Figures

Table of Contents

„Waarschijnlijk stelt dus het zenuwstelsel in het zintuigelijke, receptorische punt, zijn eischen van energieoverdracht uit den prikkel. Aan dezen eisch wordt door twee quanten lichtenergie voldaan".
ZWAARDEMAKER, 1921

"Probably it is the nervous system at the sensory receptory location that sets the condition for the energy transfer out of the stimulus. This condition is met by the energy of two light quanta".

Introduction

ZWAARDEMAKER'S insight into the minimum perceptible for vision was well ahead of his time and was not followed up until the early forties when such specific quantitative approaches to the problem of visual thresholds in terms of quantum-coincidence mechanisms were taken up again. The main contribution came from VAN DER VELDEN (1944). With ingenious considerations of a theoretical-statistical nature and some psychophysical experiments he provided a more rigid basis for the two-quantum threshold theory. The same

* Department of Medical and Physiological Physics, State University Utrecht, The Netherlands.

approach with originally identical conclusions was followed by BAUMGARDT (1950). Less complete analyses had been published already by HECHT, SHLAER, and PIRENNE (1942) and by DE VRIES (1943). We have presented elsewhere some critical remarks concerning these analyses (BOUMAN, 1969).

The quantum coincidence theory contains the idea of a threshold criterion that is based either on sampling by a number of independent discrete retinal perceptive units or on the interaction between the effects of quantum hits in individual receptors. In the latter, each receptor is considered to be the central element in its own perceptive field.

Mathematical analysis of the concept leads for both alternatives to an almost similar dependence of threshold energy E_{th} on target area a, exposure duration t and number k as the minimum number of coincident quanta required by each of the sampling units, or required in the interaction mentioned for retinal stimulation. For a and t both sufficiently large this is

$$E_{th} \sim f k \left(\frac{a t}{A T}\right)^{k-1/k}, \tag{1}$$

where f is the fraction of the incident light that is absorbed in the photopigment, and A and T are respectively area of sampling unit or of summative interaction and summation time. For $a \leqq A$ and $t \leqq T$

$$E_{th} \sim f k. \tag{2}$$

The probability W for a coincidence of at least k quanta depends on stimulus energy N:

$$W(\overline{N}, k) = 1 - e^{-f\overline{N}} \sum_{p=0}^{k-1} \frac{(fN)^p}{p!}. \tag{3}$$

This relation is only definitely true either for test stimuli whose repetitions always enter completely within a single perceptive unit when these units are distinct elements or for test stimuli whose dimensions do not exceed the shortest pathway along which receptors can mutually interact.

Experimental work and theoretical analysis suggested the validity of a twofold coincidence mechanism $k=2$ for all locations on the dark-adapted retina, including the fovea, irrespective of what type(s) of receptors were activated by the stimulus (BOUMAN, 1950, 1955, 1961, 1969).

Accepting such a hypothesis as a realistic mechanism for threshold vision, further exploration of the distinction between achromatic and chromatic perceptions leads to interesting possibilities of retinal organization: double or multiple coincidences of singly hit red and/or green cones and/or rods within a distinct group of these receptors produce achromatic or scotopic perceptions. Such a group should contain two or three red, two or three green cones, one blue cone and an eccentricity-dependent number of rods. The model predicts that color will appear in the perception when red and/or green cones and/or rods are individually hit by two or more quanta. Blue might enter the perception either by any single hit in a blue cone or by multiple coincidences of either singly or multiply hit blue cones. In this way it is possible to arrive at a model of the retina that is organized in the manner of ommatidia. The various stations along this route are elucidated below. The basis for this study is provided by a restricted number of facts, all psychophysical in nature and referring to threshold vision only. The model itself depends on arguments of a bio-mathematical nature. In the final sections some connections are established with relevant histological and electrophysiological data.

Scotopic and Photopic Cone Vision

One of the pillars of our knowledge of the primate's visual system is the dual nature of the retina: the existence of both a photopic system and a scotopic system. This duality becomes apparent in the fast and slow phases of adaptive behavior of thresholds, in a shift of the wavelength of maximum spectral sensitivity towards shorter wavelengths with decreasing illuminance, and in the colored and uncolored perceptions of stimuli. These facts are well established and their frequently experimental confirmation has confirmed this duality beyond any doubt.

However, as regards the widely accepted explanation of the supposed existence of two different classes of receptors, rods and cones, that are thought to be responsible for the two types or categories of visual perceptions, the case is different. Their existence is not a fact; it is theoretical and morphologically far from evident (Pedler, 1965).

A well-known psychophysical phenomenon contradicts the rod-cone idea as basis for the dual nature of the retina: this is the achromatic perceptions of weak monochromatic stimuli that are presented foveally. In fact, these foveal achromatic perceptions are—as introspection tells us—no different from peripheral scotopic ones. Consequently, visual theory must take account of scotopic perceptions mediated by foveal receptors. This naturally leads to the question as to whether these receptors constitute a foveal scotopic system, that is, apart from the photopic system, or whether the separation between these systems does occur at receptor level at all. If not, the same receptors can participate in photopic as well as scotopic perception.

A rather extensive study of statistical phenomena at the threshold of vision and how they relate to the quantum statistics of the stimuli yielded relevant data on this achromatic zone, or photochromatic interval. This zone or interval is the ratio between the chromatic threshold and the visual threshold. In the fovea it was found to be about 1.4 for short, small monochromatic, circular flashes, except for regions in the spectrum around 580 and around

500 mμ where this ratio was somewhat larger (BOUMAN, WALRAVEN, 1957a, b; WALRAVEN, 1962).

Recently, further studies on this subject have appeared in the literature (GRAHAM, HSIA, 1969; CONNORS, 1969). The results of various authors differ but, considering the difficulty of the task, these differences are relatively small. There is surprisingly good agreement between most studies on the constancy of the achromatic zone for wavelengths longer than 600 mμ. This parallelism between the wavelength dependence of visual and chromatic thresholds suggests that the action spectra of the receptors that contribute to these types of perception are equal. In our experiments the photochromatic interval was also measured for various choices of area and duration of test

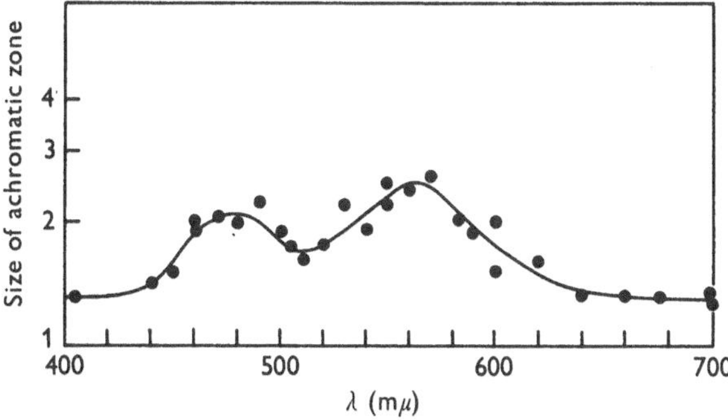

Fig. 1. The achromatic zone as a ratio of color and absolute threshold at large test stimulus diameter (60′) as a function of wavelength, for the dark-adapted fovea. (From BOUMAN and WALRAVEN, 1957a; WALRAVEN, 1962)

stimuli. A striking and basic result proved to be that both thresholds behaved similarly in their dependence on these spatial and temporal parameters of the stimuli. These findings compel us to consider the possibility that a particular receptor population, especially the red "cones", does indeed have a chromatic and an achromatic transmission system connected with it. Both systems should show threshold mechanism that are alike in some aspects but differ in others. We arrive here at a point where the twofold coincidence theory for the visual threshold naturally leads to a very specific model of retinal mechanisms and of nervous network organization.

Within the region in which the two quantum absorptions required by the theory have to occur the threshold luminance should be inversely proportional to the stimulus area; beyond this limit it should be inversely proportional to the square root of the stimulus area [RICCO's (1877) and PIPER's (1903) laws, $k=2$ in formulae (2) and (1)]. The underlying assumption which enables PIPER's law to be explained as a consequence of the twofold coincidence mechanism is that the retinal area involved can be regarded as morphologically and physiologically homogeneous. Two avoid disrupting the line of reasoning, this will be discussed in another section. In our opinion, the condition of retinal homogeneity is not too critical in those instances—including the one cited below—where we applied the two-quantum analysis.

Because the visual and chromatic thresholds both follow PIPER'S law, it is suggested that both are the result of a two-quantum coincidence mechanism.

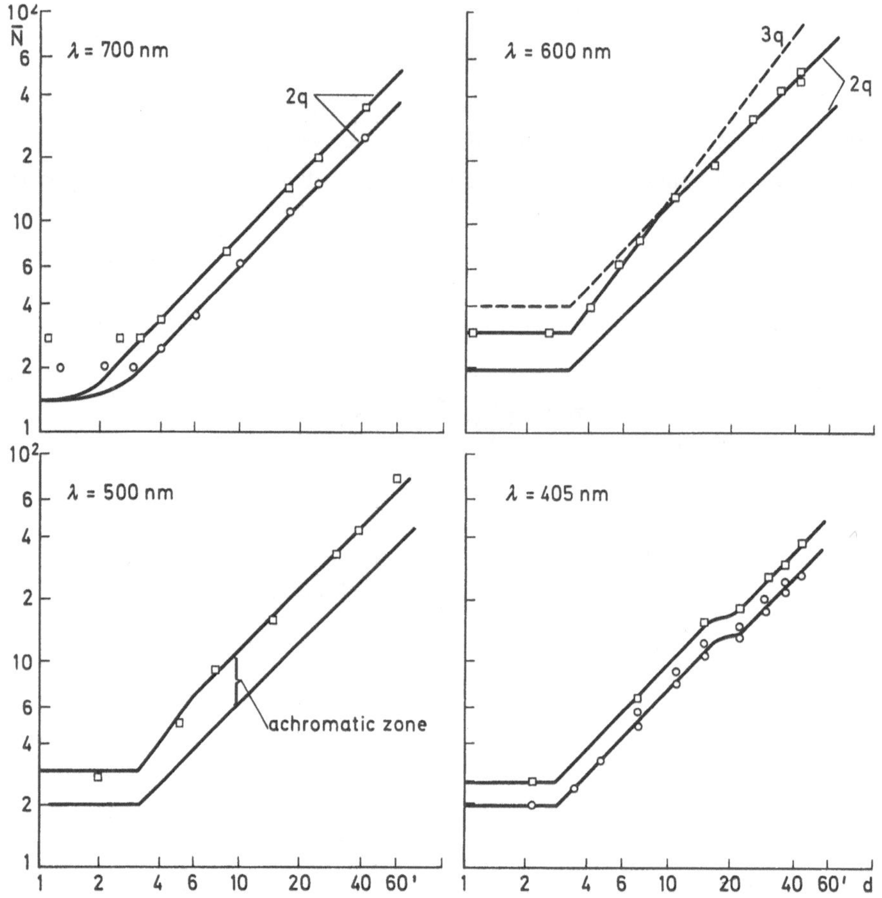

Fig. 2. Threshold for the dark-adapted fovea as a function of diameter (d) of the test flash at the indicated wavelengths. The lower curves represent the absolute threshold, the upper curves the color threshold. The curves corresponding to a two-quantum mechanism are indicated by 2 q, those corresponding to a three-quantum mechanism by 3 q. The curves for 700 mμ follow the hypothetical course when corrected for diffraction and optical imperfections. (From BOUMAN and WALRAVEN, 1957b; WALRAVEN, 1962.) Quantum numbers $\bar{N}$ on relative scales

For the higher of these two thresholds, more strictly spatial and temporal conditions should be operative. The achromatic zone of 1.4 mentioned for short flashes probably means that the area within which the coincidence must occur for the photopic mechanism is half ($1/1.4^2$) the area for the scotopic one. Unfortunately, due to blur and imperfections of the eye optics, this difference is not reflected in different sizes of RICCO's region in the area-threshold relations for both thresholds. This implies that, even for the scotopic signals, the coincidence area is less than 2 minutes of arc in diameter. It is hard to ascertain in a direct way exactly how many receptors are covered by these coincidence areas. However, the smallest possible coincidence area is the individual receptor. The consequences of this extreme case are discussed in the next section.

Human Ommatidia

The simplicity of this case and its structural and functional consequences make it an obvious and possibly correct choice. Single hits in a couple of receptors should lead to a scotopic response, and double hits in one receptor to photopic perceptions. The case where, not two, but at least five to seven quantum hits (HECHT et al., 1942) are required before light is perceived does not offer such a simple picture.

In the spectrum the photopic responses beyond 680 mμ are always red. This is understandable when we realise that for these wavelengths the absorption of the green system is negligible compared to that of the red system. Consequently, while the scotopic coincidence area is twice as large as the photopic one, mutual interaction between red cones in the scotopic system should be restricted to couples.

The achromatic zone reaches a maximum value where the action spectra of separate receptors of different systems are supposed to cross. There the chance of absorption in neighboring receptors—of the same or a different type—is maximal compared to the chance of a double hit in one receptor. We found the maximum value of the achromatic zone for normal trichromates to be about 2.25 at a wavelength of around 560 mμ.

Again, for this wavelength, PIPER's law holds true for large test areas, for the visual or absolute as well as for the photopic or chromatic threshold.

Consequently, we suggest that at 560 mμ RICCO's area for scotopic or achromatic perceptions is 4 to 5 times larger than for photopic or chromatic ones. At this wavelength the blue cones do not significantly interfere. Allowing for experimental error, our next tentative conclusion is that there are functional groups consisting of two or three red and two or three green cones. Such a group constitutes a unit within which interaction occurs between the subliminal single quantum absorptions. Such interaction feeds the scotopic pathways of the visual system. The group organization of receptors as described above fits better in a picture of the retina with distinct sampling units than a system where each receptor is the spatial centre of its own perceptive field.

For wavelengths below 500 mμ some complications are met with. The second maximum near 500 mμ in the achromatic zone (see Fig. 1 where data are given for a test stimulus of 60 minutes of arc diameter) might be due to inhomogeneity of the test area. Possibly the rods participate here. The low threshold levels for blue wavelengths and field diameters beyond about 20 minutes of arc, as apparent in Fig. 2, are due either to the generally assumed lower density of blue cones in the center of the fovea, or possibly also to the rod interference mentioned above. More will be said on this point in another section.

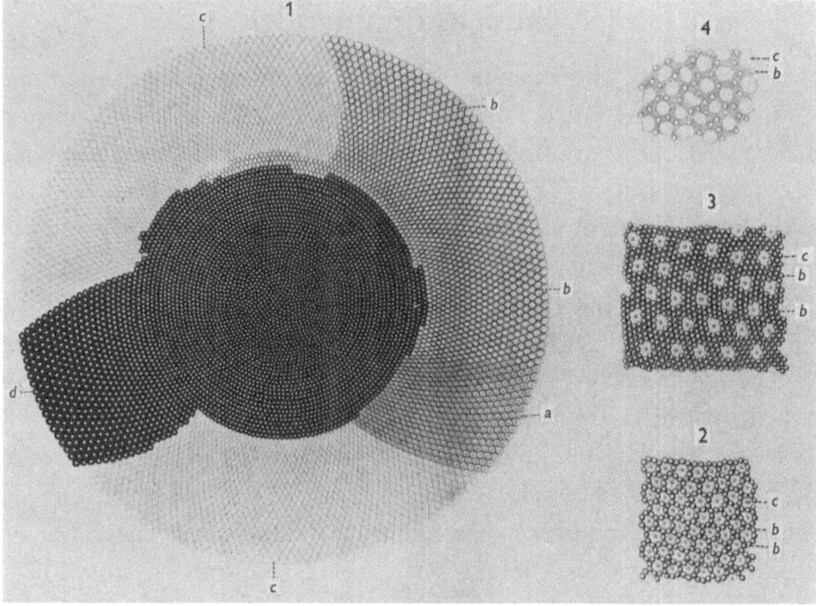

Fig. 3. Receptor mosaics of human retina, enlargement about 500; *1* fovea, *a*, *b*, *c*, and *d* indicate different depths in the preparation; *2* parafovea; *3* periphery; *4* ora serrata. In parts *2*, *3*, and *4* *b* is a rod and *c* a cone element. (From Schultze, 1866)

On the basis of this discussion we postulated (Bouman, 1969) that the cones of the human retina are organized in distinct units which may well be called "human ommatidia". The analogy with the retinular cell aggregate of the compound eye is evident. Additional clues suggesting that the retina may be organized in the manner of ommatidia are available in the nineteenth-century histology of the human retina. Schultze (1866) found evidence of a magnificent periodic mosaic pattern in the retina of vertebrates, including man.

More recently Pedler (1965) suggested that possibly Nature's grand design converged, for insects as well as for primates, in ommatidium-structured color seeing eyes. Boycott and Dowling (1969) discussed the possibility that in the primate retina each flat bipolar is in synaptic relationship to a distinct group of 6 or 7 cones, such that possibly each cone contacts only one flat bipolar

Patterns of autofluorescence of goldfish, frog and gecko retinae clearly show non-random distributions of receptors of various types over the retinal mosaic (LIEBMAN and LEIGH, 1969).

Thus, SCHULTZE's periodic receptor mosaic may be connected with a nerve network that shows corresponding periodic features. At receptor level, this implies that current knowledge of electro-magnetic wave theory and of the properties of antenna arrays becomes highly relevant for an understanding of the directional properties of the retinal receptors, including action and reflection spectra in dependence on the coherence qualities of the incident light. An anomalous pattern of essentially normal receptors in the geometric structure of the retinal mosaic could provide a possible basis for anomalous color vision. Optical coupling between receptors in such a retinal array might result in peculiar dependences of the receptor action spectra on directional parameters of the incident light waves and of the mosaic pattern (BOUMAN, 1969). For an understanding of the functions of the optical system including the servo link in the pupil and accommodation mechanism, considerations of the coherence of light waves may be as important as the energetic aspects of the light (BOUMAN, 1969).

Retinal Color Signals

The lower the light level, the smaller the stimulus size and the shorter the presentation, so the poorer the performance of the color vision mechanism will become. This is evident from color naming for perceptions around threshold level of the weak, short and small light flashes used for the study of the achromatic zone mentioned above. For instance, yellow may be called green and blue may be called red (BOUMAN, 1950; BOUMAN, WALRAVEN, 1957b; WALRAVEN, 1962). Experiments have been made to verify whether "wrong red perceptions" for, say, "green wavelengths" were different from "correct red perceptions" for "red wavelengths". They were not. This means that both red perceptions were due to the fact that only the red cones were stimulated at both wavelengths. This confirms the implicit assumption we made for our quantum coincidence analysis that led us to the ommatidium idea for the retina.

Since at the extreme red end of the spectrum a wrong chroma was never perceived, apart from scotopic perceptions, it seems safe to assume that multiple hits per red cone produce red signals in the retina. The rather symmetrical appearance of the curves for red and green perceptions around 580 mμ in Fig. 5 then suggests that green signals result when multiple hits occur in a green cone. At 580 mμ the achromatic zone reaches about the largest value and, in addition, white and yellow perceptions are more frequent than in any other wavelength region. So it would seem that white and yellow

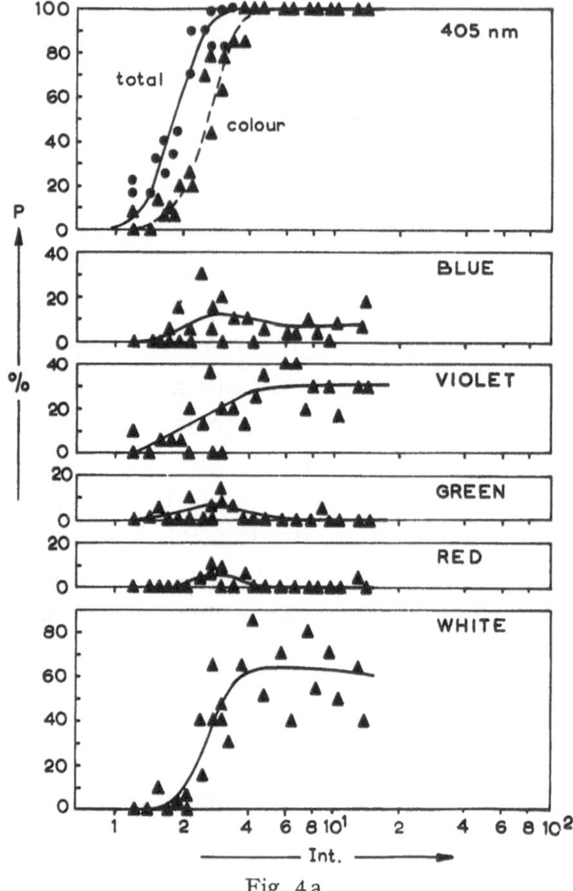

Fig. 4a

Fig. 4. Frequency of seeing curves as a function of intensity at the indicated wavelengths, for flashes of small visual angle (3′) and short duration (0.05 sec). The upper graph represents the frequency of seeing at all (total) and the frequency of seeing color (color). The other graphs represent the frequencies of seeing of the particular colors indicated. (From WALRAVEN and BOUMAN, 1957b; WALRAVEN, 1962.) The measurements were performed on the dark-adapted fovea

signals originate from the coincidence of equal numbers of red and green signals, i.e. multiple hits on individual red and green cones. Around threshold levels yellow perceptions are less frequent than white ones. Consequently the coincidence condition for yellow for these short flashes must be stricter than for white. Now the smallest unit that is larger than the receptor is the ommatidium. Rather good quantitative agreement of the observed frequencies of perceived color qualities is obtained when the hypothesis is made that yellow perceptions are produced by coincidence of equal numbers of red and green signals in a single ommatidium, and white perceptions from such a balanced red-green stimulation in a group of 5 to 7 ommatidia. Hence, when an actual test flash produces a red signal in one ommatidium and a green one in a neighboring ommatidium, white is perceived.

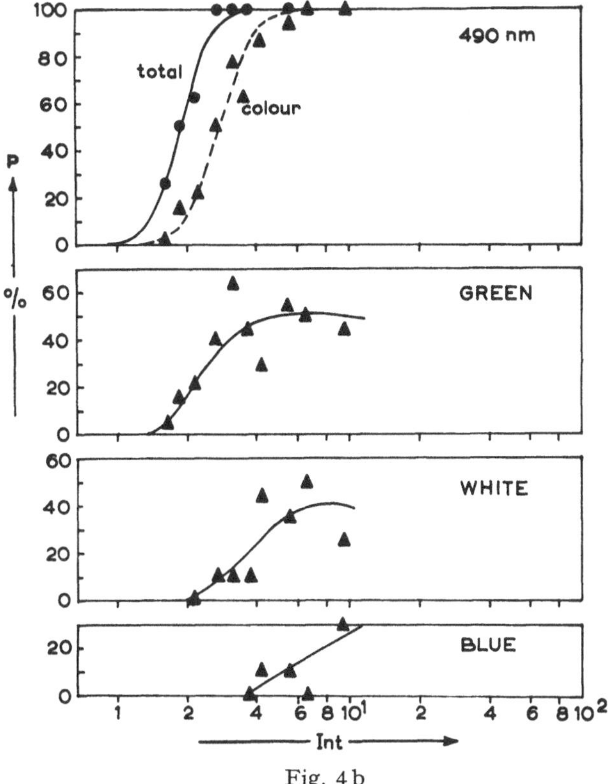

Fig. 4 b

Where the ommatidia involved in this balanced red-green stimulation are too far apart, separate red and green spots are seen. Consequently the group of 5–7 ommatidia mentioned represents the perceptive unit. Neighboring units may overlap such that each ommatidium belongs to 5–7 units. The white perceptions so far discussed occur by chance at relatively low stimulus levels. When the luminance is increased, balanced red-green stimulation always occurs simultaneously at ommatidium and at perceptive unit level. This agrees with the experience that white and yellow are perceptually closer to each other at higher brightnesses.

Fulfillment of the yellow condition in the ommatidia includes fulfillment of the white condition of the perceptive unit. The reverse is not true. Such a peculiar relationship between conditions for yellow and white might account for the white image that is perceived when a red stimulus presented to the left eye and a green stimulus to the right eye are binocularly fused. Perhaps we may now conclude that at least the retinal perceptive units are individually represented in the central visual pathways where both retinae interact.

In an earlier section we mentioned the problematic situation with regard to the analysis of the experimental results in the blue part of the spectrum. In addition, blue point-flashes are seldom seen as blue, whereas blue has an

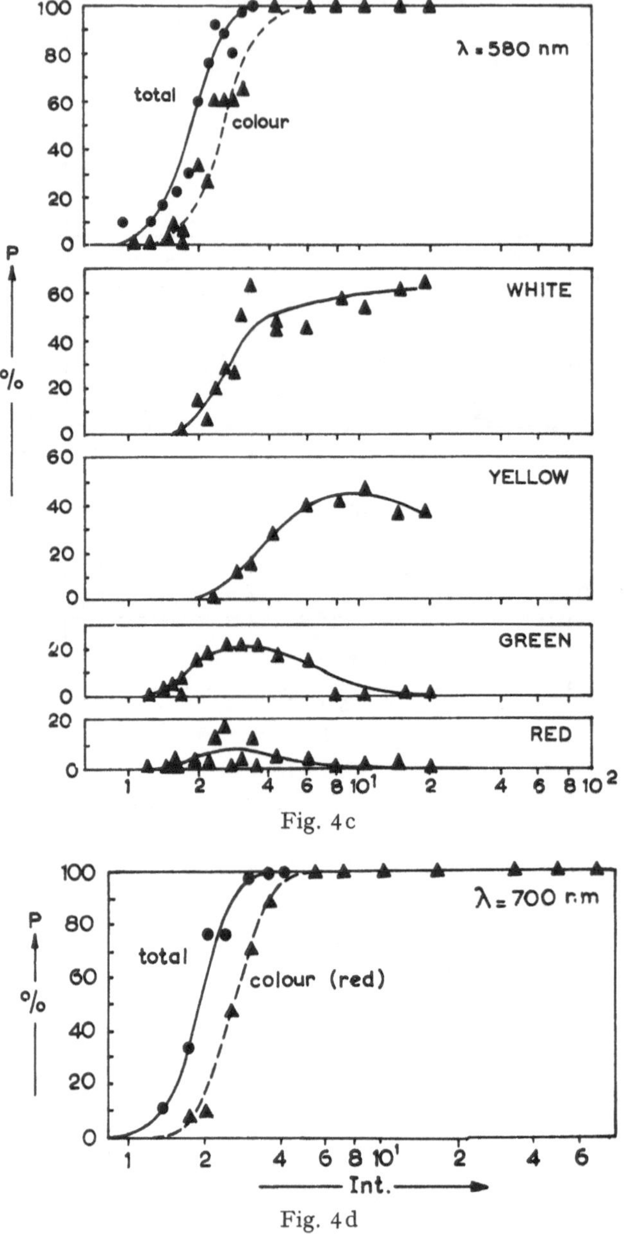

Fig. 4c

Fig. 4d

enormous color valence when presented over large areas (WALRAVEN, 1962). In the red no such difference occurs in the perception of test-fields of large and small sizes. The large variation in color naming of 405 mμ light is reasonably explained by the hypothesis that the sensitivity of the blue system is about equal to the combined sensitivities of the red and green systems. Photopic answers for large 405 mμ test stimuli, however, are always violet at any energy level. This can be compared with the red end of the spectrum where

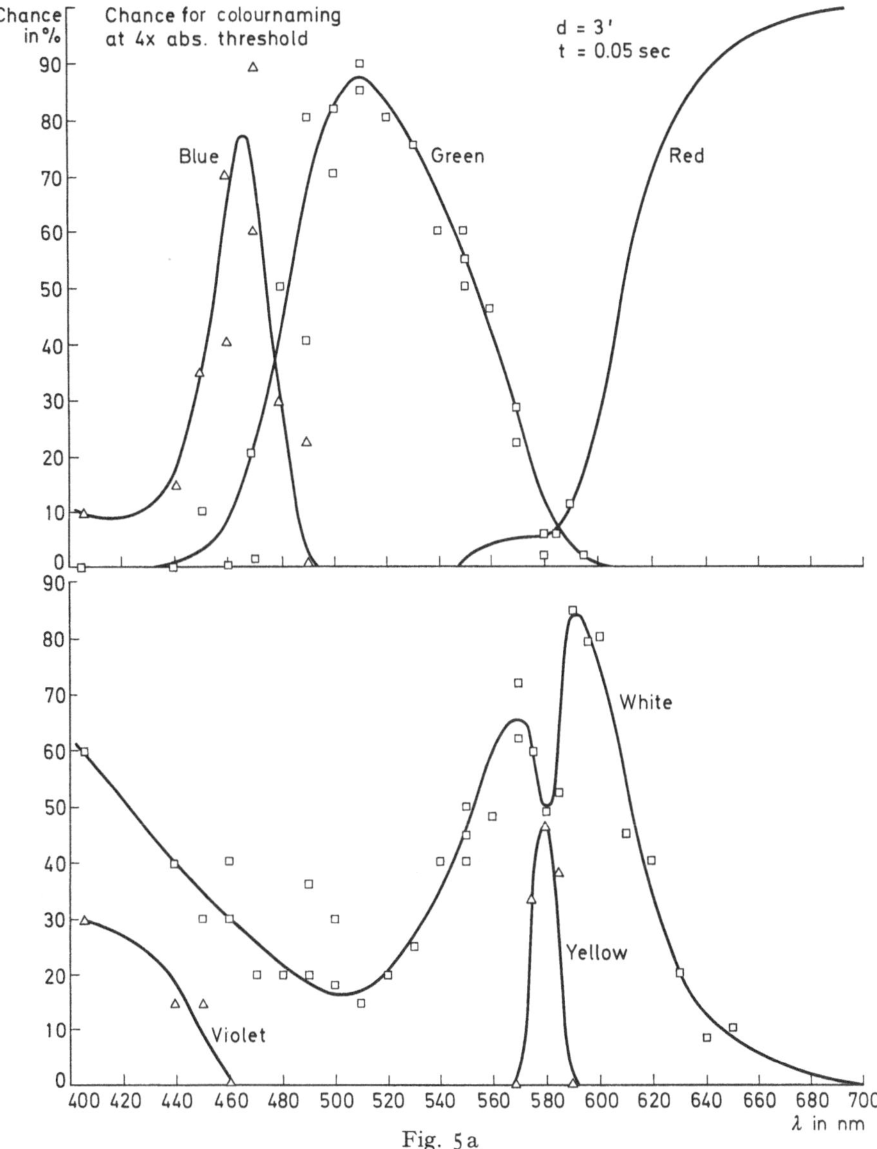

Fig. 5a

Fig. 5. The frequencies of seeing particular colors, at the threshold of color and at 4 × absolute threshold of the dark-adapted fovea for small ($d = 3'$) and short ($t = 0.5$ sec) flashes. (From WALRAVEN, 1962)

a similar situation exists for red answers. It is not likely that this peculiar difference in color naming of large and small test-flashes at the blue end of the spectrum is only due to a sparse distribution of blue cones. The frequency at which blue answers are given at low levels is still too high for such an explanation, even for the smallest size of test area. Because of these blue perceptions the central fovea cannot be said to be blue-blind. All this points to a foveal organization in which the spatial coincidence requirements for

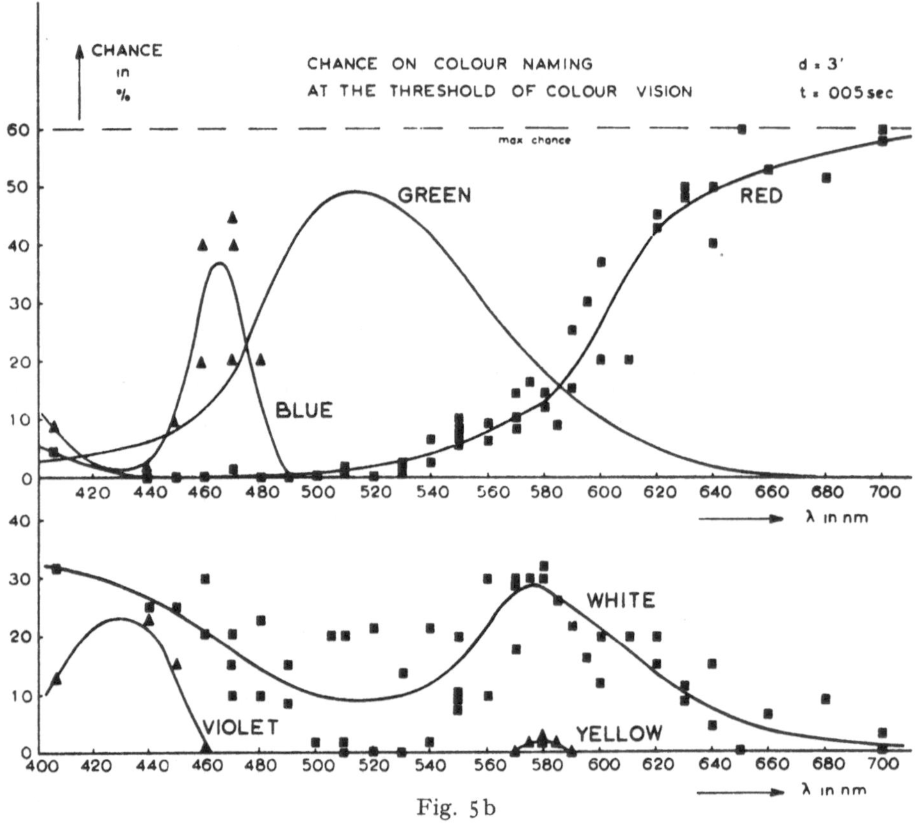

Fig. 5b

blue signals are less strict than for any other color signal, including yellow and white. We venture to make a rather speculative extrapolation: blue signals are produced by multiple coincidence within the perceptive unit of single quantum hits in blue cones, or even by singly hit blue cones. It must be stressed here that in this rather tentative model blue signal generation does not satisfy the conditions for visibility of the test stimulus. Such generation must be shared by a multiple quantum coincidence in other receptors of any of the unit's ommatidia in order to lead to a perception. In our theory, yellow signals produced by the ommatidia of a particular perceptive unit are turned into white ones by an equal number of blue signals, no matter how the latter are distributed over the unit. One blue cone per ommatidium would suffice to produce such behavior. The increasing frequency of white perceptions with decreasing wavelengths in the blue-violet region of the spectrum at low stimulus levels is possibly due to balanced red-green unit stimulation in addition to the white that results from the above mentioned combination of yellow ommatidium stimulation and blue signals. Consequently, we arrive at ommatidia consisting of two to three red, two to three green and one blue cone. This is in the central fovea. It is tempting to consider now into which channel the rod response might fall.

Photopic and Scotopic Rod Vision

The conclusions mentioned in the previous paragraph would be unacceptable if the scotopic cone and rod signals could not meet in a common system. More precisely, a single rod and a neighboring (red or green) cone both stimulated by a single quantum absorption should result in a perception.

From experiments with test flashes consisting of both red and green monochromatic light, which in the dark-adapted peripheral retina stimulate selec-

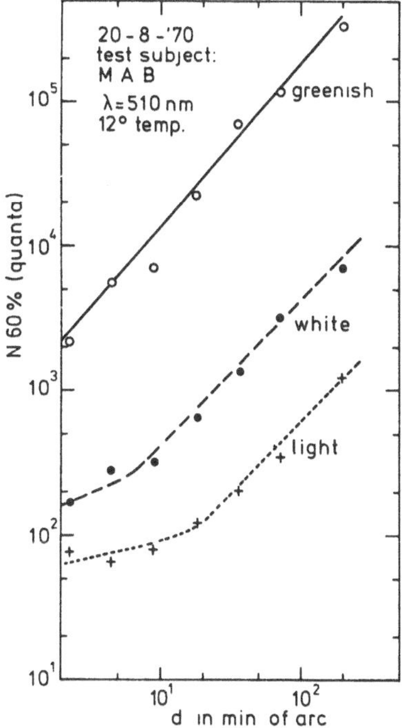

Fig. 6. Absolute, white and greenish threshold energies for dark-adapted peripheral vision 12 degrees temporal, 510 mμ, as a function of target diameter *d*. (From BOUMAN and WALRAVEN, in press)

tively red cones and rods, we know that subliminal effects in both receptor populations interact as if they were part of one system. Independent evidence for mutual interaction between red and green cones in the scotopic system came from similar studies in the fovea (BOUMAN, 1950).

In this section the main problem is whether multiple hits per rod could also result in photopic signals. The data given in Fig. 6 on the visual threshold and the chromatic threshold are similar to those of Fig. 2, but now for short test flashes 12 degrees temporal from the fovea in the dark-adapted eye for 510 mμ light. Application of the same analysis as was explored in the previous sections leads to the following chain of possible conclusions: 1) The photochromatic interval for large test areas is about a factor 8. 2) Both the visual and the chromatic thresholds are the result of a twofold coincidence threshold mechanism, because Piper's law is valid for both. 3) The spatial coincidence requirements for the visual threshold subtend an area that is about $8^2 \infty 65$ times as large as for the photopic threshold. In contradistinction to the fovea, some of this difference also appears in different Ricco areas in the threshold-area relations. 4) Around the chromatic threshold level the photopic signals must be due to rods because these are the only receptors that are stimulated. 5) Conclusion 4 is confirmed by the constant photochromatic interval found between 450 and 550 mμ (450:7.1, 510:8.0, 550:7.8). 6) The photopic perceptions are always white so that a multiple hit in a rod produces a white signal. 7) At this retinal location the scotopic system might operate in an ommatidium consisting of 2 to 3 red, 2 to 3 green, 1 blue cone and about 65 rods. Because the achromatic zone for 700 mμ is about the same size all over the retina, we believe that all ommatidia should contain this identical group of a few red cones, a few green cones and one blue cone. 8) Since the zone for greenish perceptions depends on wavelength, receptors other than rods are responsible for green signals (450:14, 510:200, 550:250).

With extended stimuli, the photochromatic interval is in our model proportional to the square root of the number of receptors of the ommatidium that can be stimulated by the type of light used for the test-flashes.

In the picture presented in the previous sections, two-fold quantum coincidence mechanisms mark the distinction between visible and invisible stimuli and between chromatic (photopic) and achromatic (scotopic) perceptions. Because of its surprising simplicity and intriguingly precise and direct specification, it is an elegant and interesting picture. However, its basis is considered to be controversial, particularly by Pirenne and Marriot (1955) and by Brindley (1963). Barlow (1956) accepted the objections raised by these authors and suggested an alternative model for threshold vision. It is based on an operationally defined imaginary luminance which is supposed to represent the influence of rest activity in the visual system. This concept of intrinsic noise or dark light was thought to be consistent with some data on frequency-of-seeing curves. In such a curve the frequency of seeing W is plotted linearly against the logarithm of the intensity of test-flashes. The concept was shown to be inconsistent with the behavior of temporal and spatial summation as apparent in Piper's law and thresholds for moving point sources, interalia, and therefore is not worth further consideration (Bouman, 1961).

Pirenne and Marriot's as well as Brindley's objections include the claim that when a k quantum coincidence concept is valid, whatever the dimensions of the test stimulus, no proper frequency-of-seeing curve can show a steeper increase with log intensity than that corresponding to results of simple application of the relation (3) for a fixed number of independent retinal units that covers a total area corresponding to the said dimensions. This claim is probably correct where each individual receptor is the spatial center of a retinal area in which the k-fold coincidence has to occur. In such a retinal organization there should be as many excitation centers as there are receptors. Indeed, many studies of the visual system assume no convergence in number from receptor towards these centers. However, it seems very likely that such convergence does occur, and if it does, the claim is false. When the test flash area is, for instance, equal to the area of the distinct coincidence unit, both being circular, and the flash contains an average of $\overline{N}$ quanta, it is trivial that there is no unit in which the average number of quanta is equal to $\overline{N}$. The actual numbers in the units for particular flash presentations not only obey the Poisson statistics of the quanta but indeed also depend on the position of the retinal unit mosaic relative to the image of the visual target. Moreover, due to involuntary eye movements, this position will differ in repetitions of the test flash. Their mathematical approach offended against this triviality. Especially for test flash dimensions that are about equal to the diameter of the distinct excitation centers, frequency-of-seeing curves will indeed show steeper slopes than application of the relation (3) for a single unit predicts. In HECHT, SHLEAR and PIRENNE'S paper (1942), for 10 minutes of arc diameter flashes, slopes appear for which application of (3) leads to k between 5 and 8. On the basis of the analyses given above, these numbers are not a conclusive argument against a lower true k-value (BOUMAN, VAN DER VELDEN, 1947). Other restrictions for the applicability of relation (3) that are more psychological in nature do exist (BOUMAN, 1969).

The origin of the controversion of the authors mentioned is their assumption that under the conditions under which their and HECHT, SHLAER and PIRENNE'S published curves were obtained, such psychological complications were excluded. PIRENNE and MARRIOT (1955) further implicitly suggest that if k were two, HECHT, SHLAER and PIRENNE could never have found slopes for which k is between 5 and 8. However, they admit that in the original study shallower slopes were also found. These were not published because their shallowness was attributed to "fatigue" and the consequent unreliability of the test subject. They further suggest that "unsufficient" randomisation in the presentation of the test stimuli and the consequent introduction of psychological artefacts did make some other authors find curves that were steeper than $k=5$–8 requires. Whatever the scientific basis of the psychologically very complex concept of fatigue, "sufficient" randomisation and avoidance of fatigue seem conflicting requirements. It is difficult to see how any author can consider himself justified to conclude that his compromise in this conflict leads exactly to the "true" steepness of the frequency of seeing curve.

Recently it was clearly shown by SAKITT (1971) that spatial summation at absolute threshold is dependent on the configuration of the test stimulus. Indeed, actual relations between quantum inputs of neighboring ommatidia in our model depend on configuration: for instance such relations for a group of four point sources on the edges of a square with $d = {}^1/_2 D$ differ from those for a circular target with area ${}^1/_4 D^2$. This includes differences in the slope of corresponding frequency-of-seeing curves, as SAKITT demonstrated. Moreover, he concluded from his experiments with multiple-spot stimuli that, if there are independent detectors that determine threshold, each having the same coincidence requirement, then that criterion must be two. His dilemma that this would not explain his data for lines of 34 minutes of arc length can probably be solved by recognition of the arguments given above in this section.

Recently it has been shown by ZACKS (1970) that around threshold level of the dark-adapted eye, a subject can, on the basis of a distinct difference in appearance, distinguish a flash of 0.03 seconds from a flash of 0.10 seconds duration. This seems to be a straightforward consequence of the difference between the Bloch times of respectively about 0.10 and 0.03 seconds of the scotopic and photopic systems.

For those locations on the retina and spectral distributions which have a photochromatic interval that is for the 0.10 seconds test flashes not larger than 2 to 3, the subject's capability to make the distinction mentioned can be a result of the fact that around threshold-level short flashes tend to stimulate the photopic system and long lasting flashes only the scotopic system. Similarly do small flashes dominantly stimulate the photopic system and extended stimuli the scotopic system as is apparent in Fig. 6. This in spite of the fact that the threshold energy as such is about the same for both types of flashes.

The conditions of the elicitation of a blue signal in any perceptive unit are not supposed to include the ommatidium's condition for visibility of the test stimulus. In the rod-free central fovea only stimulation of the red and green cones can sensibilize these blue signals. At other locations of the retina, sensibilization is aided enormously by the contribution of the rods. This might further explain the tale of the blue blind fovea. It also could make clear how some investigators have been led to regard the rods as blue receptors (MCCANN, BENTON, 1969; WILLMER, 1965; TREZONA, 1970).

General Coincidence Concept and Visual Adaptation

The validity of PIPER'S law has never been disputed where a simple dual quantum coincidence mechanism determines the threshold and the relevant properties of the retina are sufficiently homogeneous.

Yet objections have been raised concerning the validity of PIPER'S law in spite of the many cases where the experimental results seem to follow the law fairly closely. It has even been categorically denied (WALD, 1967) that any such a relation exists as $I_{th}^{n} \times \text{area} = \text{constant}$. However, WALD's own data do not deviate by more than 12% from $I_{th}^{0.4} \times \text{area} = \text{constant}$ over a range of areas of several decades. Consequently it seems appropriate to regard a twofold quantum coincidence mechanism as a fair model of threshold detection. In retrospect it seems that the retinal areas for which PIPER's law was verified could indeed be considered sufficiently homogeneous.

We tested relations (1) and (2) in a great variety of situations, including linear and circular targets as well as subsequent relations for moving point sources, for different wavelengths of the stimuli, over a large range of intensity of adaptation, both foveally and in the periphery of the retina. In the dark-adapted eye we confirmed, as already mentioned, the general validity of $k=2$. With increasing light adaptation k tends to increase. Because the accuracy with which k can be determined decreases rapidly with increasing k, no more can be said on the basis of such threshold measurements than that the order of coincidence required for vision rises above $k=8$ with increasing luminance of the adapting light (BOUMAN, 1952; BOUMAN, TEN DOESSCHATE, 1962).

We want to stress here that under stationary illumination, as well as during adaptation to new environmental conditions, the quantum coincidence mechanism explains the behavior of thresholds: their magnitudes as well as their dependence on spatial and temporal stimulus parameters. It is the only currently available "theory", which describes such visual functions as the threshold energy of a moving point source.

The eye functions over a luminance range subtending some twelve decades. It is obvious that the nervous system cannot adequately transmit the incoming visual information without some adjustment or adaptation. It is a well-known fact that variation of sensitivity is much greater for extended targets than for small ones. From the formula $E_{th} \sim f k \left(\frac{at}{AT}\right)^{k-1/k}$ it is easily seen that with larger a or longer t, E_{th} variation will become greater due to variation of k. Only if the lowest possible k-value is indeed as small as 2 can the experimental data be explained by variation of k. It is another, admittedly indirect, negative demonstration that k_0 is greater than 2–3 for the dark-adapted eye.

Next we discuss how the retina adjusts its coincidence requirements to the actual environmental conditions. So far, the analysis has been devoted to threshold dependence on the spatial and temporal attributes of the signal. We now come particularly to dependence on luminance and on the spectral energy distribution of the stimuli.

The De Vries-Rose Law and Weber's Law for Visual Discrimination

A fascinating problem in the science of vision is how the retina manages to process the incoming information and rid itself of the inherent quantum fluctuations in the signal that are irrelevant to the task of the visual system and can be considered "noise". This implies that a decreasing level of illumination obscures the where and when of the stimulus in proportion to the increased average separation in time and space of the quanta that constitute it. This sets a physical quantum-statistical limit to the possibility of detecting small differences in quality or quantity of light. Is there any indication that

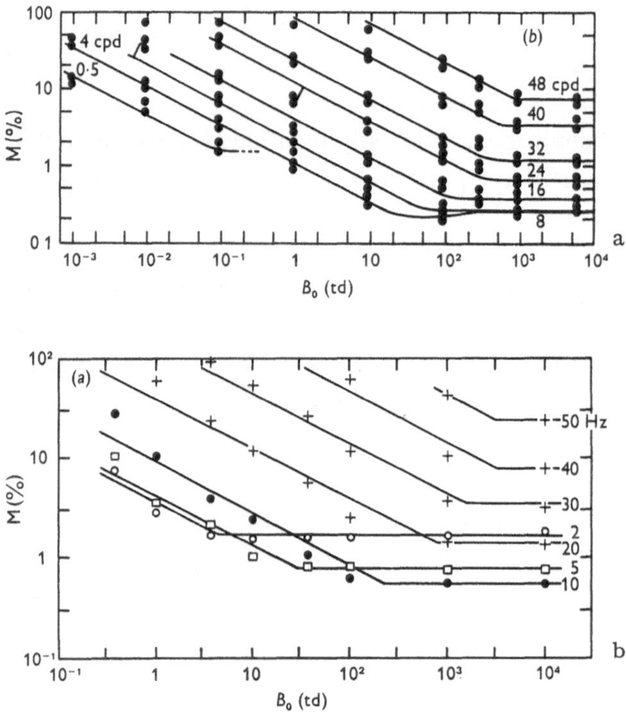

Fig. 7. Thresholds for sinusoidal modulation of luminance (M percent) as a function of average luminance level B_0 in trolands, for foveal vision and with temporal (Hz) and spatial (cycles per degree, cpd) frequencies as parameters. (From VAN NES et al., 1967)

quantum noise is indeed a limiting factor in vision? Pursuing the line of thought presented by DE VRIES (1943) and ROSE (1948), we have shown that in many instances the detection thresholds are indeed proportional to the square root of the adapting intensity B over a varying number of decades. Typical recent results are reproduced in Fig. 7 (VAN NES et al., 1967).

The experimental evidence corroborates the assumption that quantum noise is the limiting factor because the fluctuation concept predicts that brightness as well as colour discrimination will improve with $B^{\frac{1}{2}}$.

Results of the type presented in Fig. 7 have led to the prediction (BOUMAN, 1952b) that where the DE VRIES-ROSE law holds the variation in the concentration of visual pigment is negligible. This has now been shown experimentally by various authors by means of retino-reflectometry.

The validity range of the $B^{\frac{1}{2}}$ law increases with increasing eccentricity of the retinal location under investigation. This can also be seen from the results in Fig. 7, where high spatial frequency targets which can only be resolved foveally show DE VRIES-ROSE behavior over only two decades. Targets with less fine detail follow the DE VRIES-ROSE behavior over as many as four decades of background luminance.

We also studied the effect of using different colors for the test flash and adapting field. Because of their deviant behavior, we especially mention the thresholds for red test flashes on a green background in peripheral vision. In the low luminance region this background only stimulates the rods scotopically, whereas the test stimulus triggers the photopic system by multiple quantum hits per red cone, if it is even slightly above the visual threshold. We found that scotopic rod stimulation does not impede or mask perception via the photopically stimulated red cones. The reverse does not hold true, as we gathered from experiments with green flashes in a red background: photopic perceptions do mask scotopic ones.

The essential features of the coincidence concept proved to be valid for all background test stimulus relations. Thus, for discrimination of contrast, the threshold behaves like a k-fold coincidence mechanism, k being proportional to the square root of luminance level in the range where the DE VRIES-ROSE law holds (BOUMAN, 1961).

The validity of the coincidence concept and the variation of its k-value implies here, as it does in visual adaptation more generally, that for extended and prolonged test stimuli variation of threshold is larger than for small, short flashes. This is again apparent from formula 1. The DE VRIES-ROSE square root low of k will result in a square root dependence of threshold on adapting luminance when the test stimulus area $a \leqq A$ and its duration $t \leqq T$.

The square root behavior of thresholds breaks down at low luminances as we touch upon the absolute threshold. Further decrease of background luminance gives no further decrease of threshold; k has reached its bottom value of 2. At the other end of the DE VRIES-ROSE's luminance range another law, WEBER'S law, takes over. In the WEBER range threshold values become proportional to background luminance. In Fig. 7 this behavior is apparent on the high luminance side of the ordinate. Here the performance of the eye no longer follows the detection limit as determined by quantum noise. In the DE VRIES-ROSE range the eye approaches this limit by a factor of about 3, when one presupposes an independent sampling by retinal units of the size of the coincidence areas A. In some of the following sections we suggest that thresholds show WEBER behavior when all receptors in the area involved are continuously stimulated photopically: all receptors are subject to multiple hits at least once per photopic sample period T_p. Then the fraction of the receptors that are stimulated photopically is independent of the quantum noise.

The eye, given the size of perceptive units, the duration of sampling periods and the types of receptors present, has here attained its maximum performance as determined by receptor density, at least, as far as spatial and temporal resolution is concerned. This implies that for this maximum resolution the receptors in their photopic responses do not need to deliver more than all-or-none signals.

Square Root Coincidence Scalers: De Vries-Rose Machines

An important principle in the previous section is the fluctuation concept. One assumes that the nervous system limits the flow of events carrying information about the stimulus until the noise in the message is reduced to a barely detectable level. This means that at some decisive stage of nervous transmission the output should increase with the square root of the intensity. The quantum coincidence mechanism performs this job when $\bar{n} \pm \bar{n}^{\frac{1}{2}}$ quanta are transmitted as $\bar{n}^{\frac{1}{2}} \pm 1$ events. The nervous message will then get through without a significant loss in accuracy (BOUMAN, 1963). This process is not a simple counting in groups of k. What actually could happen can become clear when such a detecting machine is constructed in the laboratory. The crucial question here is how the eye knows so easily and without elaborate statistical analysis which k-value to use for optimal signal reduction. We found a simple answer to this question: an adapting system, requiring for each successive output signal a fixed number (ϱ) of input pulses in excess of those required for the previous output pulse, produces "on line" a square root relationship between output and input. Thus $\varrho + 2\varrho + 3\varrho + \cdots + n\varrho = \frac{1}{2}\varrho n(n+1)$ input pulses produce n output pulses when ϱ input pulses are required to generate the first output pulse. Substitution of $\varrho = 2$ gives an almost perfect square root relation between the total number of input and output events since the machine was started. The mathematical principle of the system is the fact that the summation of the elements of an arithmetical series leads to a quadratic relation. Combining this procedure for escalation of the scaling factor with any procedure for decreasing this factor when the input frequency is lowered results in a "DE VRIES-ROSE machine". It is a device in which the essential feature is in effect an inhibitory feedback (BOUMAN, AMPT, 1966) and in which the scaling factor is proportional to its average output frequency.

In the further development of this concept towards a mechanistically realistic model it was recognized that an adapting coincidence mechanism can be represented functionally by an adapting leaking integrator (VAN DE GRIND *et al.*, 1971a). In the first mechanism a rectangular time window of duration T continuously samples the input. An output event is produced when k input events are covered by the window, irrespective of the interval distribution of the input events. The leaky integrator is a device that signals the moments at which its potential reaches a threshold value θ. Each input event raises the potential by a constant value taken further as one, after which, as in an *RC* circuit, the potential decreases exponentially with time constant $T = RC$. It can be shown mathematically that for $n \ll 1$ and $k-1 < \theta \leqq k$ for both mechanisms the output frequency $\nu \sim \bar{n}^k$. From this it follows that formula (1) is valid for both mechanisms. Because of its ability to vary the threshold θ continuously, the leaky integrator model no doubt

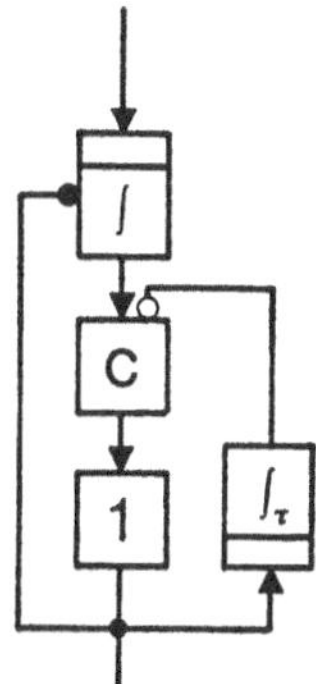

Fig. 8. Schematic diagram of De Vries-Rose machine consisting of two integrators, comparator and one-shot generator. (From VAN DE GRIND et al., 1970)

seems physiologically more realistic. The arithmetical escalation of the threshold criterion k or θ under light-adapting conditions can be realized for both mechanisms. Such escalation leads to responses of the machine not unlike on-effects in which successive intervals between output events increase arithmetically. Under dark adaptation, de-escalation of k or θ could be regulated by the statistics of some spontaneous—or rest-activity, like the regeneration of photopigment molecules. This could furnish a direct and simple explanation of the exponential decrease of psychophysical thresholds during dark adaptation (VAN DE GRIND et al., 1970, 1971a and b).

Finally, it is relevant to discuss in this section another important aspect of the proposed DE VRIES-ROSE machines: the fact that the input channels of the machine are not necessarily identical to the channels to which the results of the machine's logarithmic operation are applied.

More particularly it might very well be—and in the following we will produce evidence for it—that the actual coincidence requirements of an ommatidium for the perceptual threshold are derived from the quantum flux in all the rods and red and green cones of the whole perceptive unit in which the particular ommatidium is the central element. Such a coincidence is supposed to trigger the perception of the test stimulus. It means that we postulate lateral pathways for mutual inhibition between neighbouring ommatidia.

There are three psychophysical phenomena that would find their respective explanation in this center-surround organisation of the DE VRIES-ROSE machine. 1) If the appearance of a long-lasting and extended test flash at threshold level were entirely due to isolated cases of quantum coincidences, a fluctuating pattern of scintillations would be seen. This is not the case. Evidently such coincidences trigger a response and this response in turn seems to facilitate nervous excitation due to single quantum absorptions which would otherwise be lost (BOUMAN, 1955). This means that all ommatidia that have sent signals into the input channels of the scalers contribute to apparent area, duration, brightness and color when the stimulus is finally perceived. 2) There are many instances in which it has been shown that thresholds for long-lasting extended stimuli deviate from simple coincidence theory as expressed in formula (1). These stimuli evidently inhibit their own perception because they stimulate the square root scalers by means of the same single quantum absorptions as mentioned under 1. There is a consequent increase of threshold requirements, including the deviations mentioned (BOUMAN, 1950, 1955). 3) The eye never quite achieves the performance of an ideal detector. We define the ideal detector as a machine that conforms to the simple arithmetical adaptation procedure described above, with $\varrho = 1$. Indeed, for the perception of small and short flashes for which self-inhibition can hardly occur, ΔB is still about 4 to 5 times the quantum noise per area A and period T in the background.

This is due as suggested above under 1, to the fact that the ommatidium's coincidence conditions are determined by the quantum noise over all 5–7 ommatidia of the surrounding perceptive unit and, moreover, because of the larger area involved, over an elongated integrative period of some $3\,T$. It further safeguards the system quite effectively against false responses.

Linear Coincidence Scalers: Weber Machines

To establish the square root input-output relation by an adapting scaler, the arithmetical strategy in a feedback inhibitory process seems the only method. There is not such a singular principle for the WEBER type of threshold behavior so various types of adaptation strategies were explored. First, WEBER behavior can be obtained from an adapting scaler following the arithmetic adaptation strategy. In this case, the inhibition is applied not in a feedback but in a feedforward control loop. A simple realisation is a leaky integrator whose voltage represents the scaling factor. For a graded input variable, as well as for an input consisting of pulses with a sufficiently high repetition frequency, the average scaling factor in the steady state is proportional to the input. Consequently, the output of the scaler is independent of the input. However, because of the time needed to change the scaling factor, the system behaves as a kind of differentiator of the input signal with the time constant of the integrator as its characteristic parameter. Immediately after a sudden rise in input intensity, the scaling factor is too low and the input is divided by too small a scaling factor, which means that the output rate has increased in proportion to the step in the input. The scaling factor then starts to increase and after some time approaches the value corresponding to the new input. For an input consisting of pulses that are exponentially distributed in time the stationary output frequency $\bar{m}$ becomes proportional to the input frequency $\bar{n}$ only when $k=1$ and $\bar{n} \ll 1$. This follows from what has been said about coincidence scalers in previous sections.

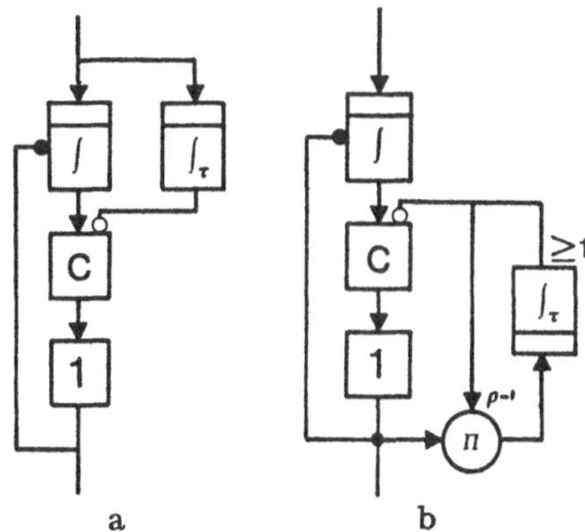

Fig. 9a. Schema of Weber machine. a: based on arithmetic adaptation strategy in feed-forward loop consisting of two integrators, comparator and one-shot generator

Fig. 9b. Weber machine similar to Fig. 9a but now based on geometric adaptation strategy in feedback loop (from KOENDERINK et al., 1970) and consisting of two integrators, comparator, one-shot generator and multiplicator

An alternative WEBER machine that seems equally realistic is based on an adapting scaler which, like the DE VRIES-ROSE machine, has a feedback control. Of course, we then have to change the arithmetic adaptation strategy. An equally simple strategy proves to produce the required input-output relation of the scaler: geometrical adaptation. An output event at time t will affect the scale factor k like $k(t+\varepsilon) = \varrho\, k(t)$, where ϱ is a real number greater than one and ε small enough to give no output pulse between t and $t+\varepsilon$. For N_o output events $N_i = 1 + \varrho + \varrho^2 + \dots + \varrho^{N_{o-1}} = (\varrho^{N_{o-1}})/\varrho - 1$ input events are needed if one takes $k_o = 1$. For not too small N_i, N_o is $\sim \frac{\ln(\varrho-1)N_i}{\ln \varrho}$. If one more output event is needed for the threshold δ, the increment threshold will be $\varrho^N = \delta(N_o)$ input pulses or $\delta(N_i) = (\varrho-1)N_t$. We write $\ln \varrho = \varrho - 1$ because we consider $\varrho - 1 \ll 1$. Proportionality of δ with N_i results (KOENDERINK et al., 1970, 1971a).

This WEBER machine shares with the DE VRIES-ROSE machine the pecularity that only information available at the output is needed by the adaptation mechanism, irrespective of the magnitude of the input level. For both types of WEBER machines, just as is the case for the DE VRIES-ROSE machine, the threshold will monotonously decrease in the absence of input events. For an exponential time-decay of the scaling factor between successive input events, a very simple input-output relationship in the stationary case is found for the alternative WEBER machine mentioned. The output tends to become constant, and does not depend at all on the input intensity as is the case with the WEBER machine with the feedforward arithmetical inhibitory mechanism.

The input event rate must be high enough to make this happen. For lower rates, the threshold will almost always be near its lowest value and effectively no adaptation takes place. The output intensity is then proportional to the input intensity, and when $k_o = 1$ the machine again acts as a single-event counter.

It can be shown that not only in the steady state, but also under dynamic input conditions like a step function, as well as in periodic and exponential time courses, both types of WEBER machine behave quite similarly in a qualitative sense. Even the DE VRIES-ROSE machine does not react very differently from the WEBER machines to such input variations. If adapting scalers are essential mechanisms in the data processing of the central nervous system, we may consequently conclude that the dynamics of supra-threshold input-output relations cannot tell us much about the essential features in this processing, unless we have accurate absolute values of and statistics for event rates at all levels in the system. This is the main reason why this discussion is largely restricted to the study of these relations around the psychophysical threshold level.

Neural Quantum Detector

All the machines just mentioned represent models that fit in to a neural quantum theory: an additional packet of input events representing the machine's detection threshold is represented in the machine's output by a single event or a fixed number of extra output events. If such a model is to be realistic, we cannot overlook the problem of the design of an additional component to detect these extra output events among a group of fixed output events which may be rather large. Such an element might work on a single output channel and compare each sample covering a period T with the preceding sample, or group of samples, or make a simultaneous comparison of an actual sample of a single machine with the output(s) of one or more neighbouring machines. The whole thing would behave like a mechanism that is expected to give a yes-response to any change in excess of a fixed number in the value of the adapting scalers. In such a picture, a test stimulus is detected through the change in the scaling factor. In other words: test stimuli are perceived only when they succed in changing the state of adaptation of the eye.

The model component concerned is easily built from an integrator and a comparator so that a positive response is produced if the number of extra output events during a time t is greater than l.

In a more general type of l-detector the mean output intensity of the preceding scaler is determined by an integrator with a long time constant $T \gg t$, the instantaneous value by an integrator with time constant t. When the difference between them exceeds l, the l-detector produces a yes-response. These machines are able to operate on both pulse trains and graded signals. Yet

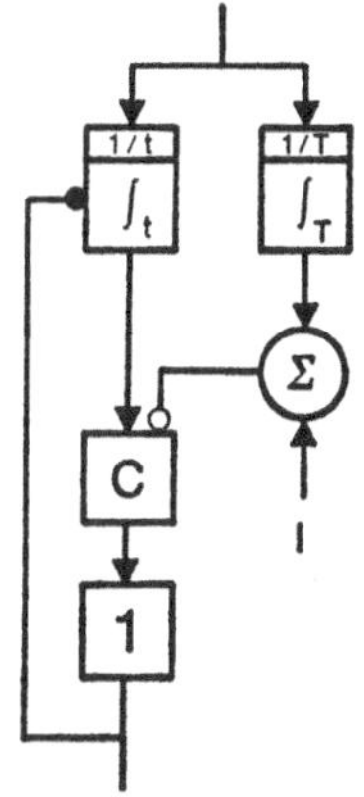

Fig. 10. Schematic diagram of general l-detector (from KOENDERINK et al., 1970) consisting of two integrators, adder, comparator, one-shot generator

another machine is formed by the introduction of a one-shot and a reset, as in the machine shown in Fig. 10. This l-detector responds with an output event as soon as the content of the integrator with the short time constant momentarily exceeds the content of the integrator with the long time constant by at least l. One can easily verify that this machine responds with several output events for suprathreshold flashes; in fact, it signals the intensity of a flash in units of the increment threshold. Consequently the machine is sensitive to successive contrast. As such, it is equivalent to the feedforward type of WEBER scaler. If the integrator with the long time constant operates, as in the center-surround arrangement of the square root scalers, on spatially channels different from the integrator with the short time constant, a further generalization is obtained. Now, in addition to detection of a temporal change, a purely spatial pattern can be detected if the outputs of the WEBER machines of two neighboring channels differ by at least l.

DE VRIES-ROSE machines—as well as WEBER machines and l-detector—can all be obtained from a few simple arrangements of control loops. The network of adapting scalers and l-detectors as discussed in these sections, no matter what their actual inhibitory strategy may be, shows output event rates which in the stationary states are independent of the event rates in the input channels.

The absence of any physically significant temporal changes makes the system lose all information about the spatial pattern, colors and intensities present at its input! Where such temporal changes are lacking, apparent hue and brightness perceived on any location of the retina becomes independent of the stimulus and retinal sensitivity at that location. This can explain why a homogeneously illuminated part of the retina is homogeneously perceived, irrespective of the large differences in retinal constitution over the area involved, even when blind spot or fovea are included. This pecularity becomes all the more apparent when the remaining temporal changes due to eye movements are eliminated by adequate stabilization. Then all perceptions fade. However, the increment thresholds remain the same under stabilized and unstabilized conditions. They are still determined by the values of the adapting scalers. These remain dependent on actual input event rates.

Mesopic Vision

In addition to photopic and scotopic vision one distinguishes mesopic vision in which both the traditional components of the duality theory are active. Although the merits of its introduction as a separate category of vision have been questioned, it has already been accepted for several deccades. The idea still does not seem very useful, but we suggest here a very accurate definition of the luminance range within which the eye sees mesopically at a particular retinal location. At the threshold of vision, perception is determined by multiple coincidences of singly hit receptors of the ommatidium. From the luminance at which these coincidences occur more than half of the time until the luminance at which multiple coincidences in individual receptors appear, vision is probably truly scotopic. The higher end of this scotopic range can be normalized as that luminance at which during half of the time at least one such multiple hit occurs in at least one receptor of the ommatidium. Another critical luminance is that at which practically all the receptors are multiply hit all the time by incident quanta. From this point on, all receptors might continuously act photopically, as we saw in preceding sections. Consequently this point marks the lower limit of truly photopic vision. In between we find the mesopic range in which the fraction of receptors that are multiply hit moves from about zero to almost one. Given the sensitivity of the receptors for the particular type of light used and the number of these receptors sensitive to this light, quantitative measures for the limits of the ranges can easily be calculated.

When more than one receptor type is sensitive to the test light, another complication arises; in the luminance range where not all types function either photopically or scotopically the ratio of the numbers of the different types that produce color signals depends on luminance. This makes the perceived hue dependent on luminance, no matter how the retina deduces its color information from the receptor outputs: either by taking ratios between numbers of color signals of the receptors, e.g. R/G, or by considering differences between these numbers, R — G for instance. This means that in this mesopic range one expects to measure the Bezold Brücke effect.

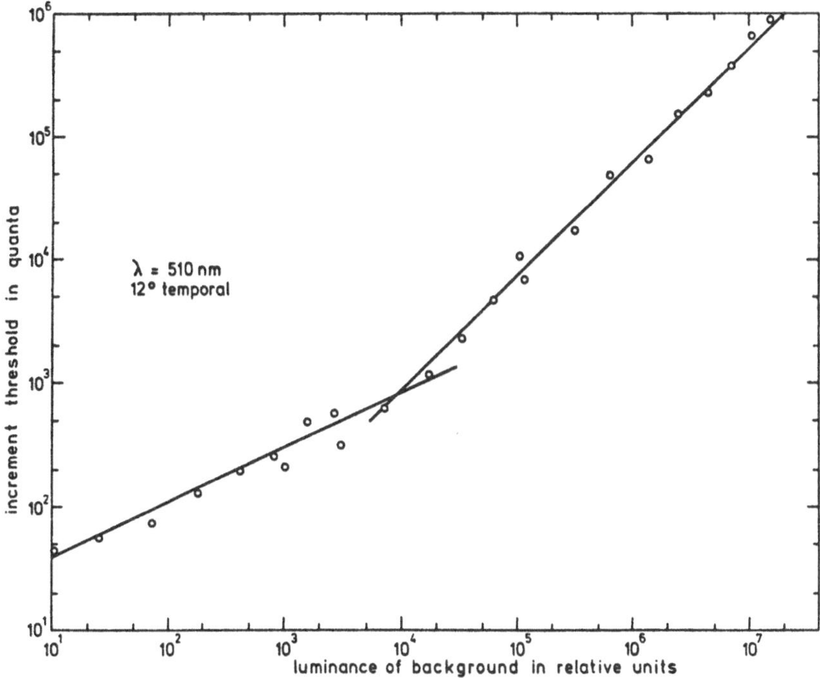

Fig. 11. Increment threshold energies for circular targets, 9′ diameter, 12 degrees eccentric from the fovea for 510 mμ light as a function of background luminance in relative units B. (From Bouman and Walraven, in press)

Naturally we must ask how the luminance ranges of de Vries-Rose and Weber refer to these three categories of vision. Fig. 11 gives increment thresholds for 510 mμ flashes of circular targets, 9 minutes of arc in diameter on a 510 mμ background 16 degrees in diameter and located 12 degrees temporal from the fovea. These results agree with the picture in which the Weber region starts at luminances where almost all the receptors function photopically. Indeed, this is the case for $\Delta B \approx 1000$ quanta, as the data from Fig. 6 show. This means that the Weber type of adaptation is connected with the production of color signals by the receptors. It thus seems that the individual receptors represent Weber machines which enter their adaptation range as soon as they are subject to multiple quantum hits.

Consequently, the data from Fig. 6 predict that the DE VRIES-ROSE luminance range will subtend a factor of about 1 000, which value is fairly confirmed in Fig. 11.

These conclusions were drawn from the following considerations: 1) DE VRIES-ROSE luminance range is proportional to the number of receptors per receptive unit; 2) the scotopic and the adjacent mesopic luminance ranges are both proportional to the square root of this number; 3) together they constitute the DE VRIES-ROSE range; 4) in our model the limiting scaling factor of the DE VRIES-ROSE machines is also proportional to the square root of the number of receptors per unit. The scaling factor of this machine determines the contrast threshold $\Delta B/B$. This means that 5) at WEBER luminance levels $B/\Delta B$ is proportional to the square root of the number of receptors per unit and is consequently also proportional to the diameter of the unit at the location concerned.

This last point has interesting consequences for the visibility of border contrast. Suppose that this contrast is beyond the value needed to stimulate the WEBER scalers of the receptors along this border and that this border is a straight line going through the center of the unit. Then the border will become visible if it is displaced in a direction perpendicular to it. The displacement needed to render the border visible is independent of the retinal location. Indeed, in any unit, no matter what its size, such a displacement of fixed amplitude will stimulate a number of receptors proportional to the diameter of the unit and hence proportional to the square root of the total number of receptors constituting the unit.

Color Discrimination

The discussions of threshold behavior in the above paragraphs have concentrated on brightness contrasts. Elsewhere (BOUMAN, 1969) we explored this model further, by introducing color channels in which the color signals of the receptors in the perceptive units are further processed. Suggested inputs for these channels were $(R-G)$ and $(R+G)$—B where R, G and B are respectively the color signals of the red, green and blue cones.

It was proposed that the threshold criterion in the red-green channel corresponds to that in the luminance channel. This suggestion originated from the validity of the DE VRIES-ROSE law for color discrimination. Indeed, in the range of luminances for which this law holds for brightness discrimination, the thresholds for color differences are also determined by the quantum noise (VAN DER HORST, BOUMAN, 1969). This means that the input in the red/green color channel is scaled by the same factor as the input of the brightness channel:

$$\{r R_o + g G_o + p P_0\}^{\frac{1}{2}} \tag{4}$$

in which r, g and p are the number of red and green cones and of rods in the unit, and R_o, G_o and P_o are the number of quanta absorbed per T photopic per red or green cone or rod. When we include the WEBER luminance range, the scaling factor becomes approximately

$$\{r(1-e^{-R_o})+g(1-e^{-G_o})+p(1-e^{-P_o})\}\ \{1+R_0^2+G_0^2+P_0^2\}^{\frac{1}{2}}.$$

For $R_o \cdot G_o \cdot P_o \ll 1$ formula (5) is the same as formula 4, for $R_o \cdot G_o \cdot P_o > 1$ formula (5) can be approximated by $(r+g+p)\ (R_0^2+G_0^2+P_0^2)^{\frac{1}{2}}$. More eccentrically the color channels are congested by the white signals of the rods. However, brightness discrimination profits from the increased receptor population per unit at these peripheral locations. Consequently, brightness and color discrimination functions have a complementary distribution over the retina.

It is rather speculative to go further into the question of whether blue signals in the yellow-blue channel are masked by the other cones and the rods. Anyhow the blue signals might be relatively less subject to quantum noise than would be supposed from the ratio between the number of blue cones and the number of other cones and rods per unit area. Consequently, in the DE VRIES-ROSE luminance range the weight of the blue signals in hue discrimination will ostensibly increase with luminance, as was found earlier in psychophysical experiments (WALRAVEN, BOUMAN 1966).

Quantal Effects in the Electrophysiology of Vision

Since the early' twenties there has been no longer any doubt about the fact that the receptor can be activated by the absorption of a single quantum of light. The evidence was provided by psychophysical methods. Only recently did the state of the art in electrophysiology reach a comparable level of sophistication. HAGINS et al. (1970) demonstrated that in darkness a steady current flows through the plasma membrane of the outer segment. This current is balanced by an equal outward current that is distributed along the remainder of each rod.

Flashes of light produce a photocurrent which transiently reduces the dark current in the isolated rat retina. The photocurrent is produced by a local action of light within 12 μm of its point of absorption in the outer segment. The quantum current gain is greater than 10^6. The electrical space constant of rat rods is greater than 25 μm, so that the electrical effects of the photocurrent are large enough at the rod synapses to permit single absorbed photons to be detected by the system. HAGINS et al. investigated the relevant voltages, currents and resistances of the receptor layer with arrays of micropipette electrodes inserted under direct visual observations by infrared microscopy.

Before HAGINS' et al. made their study, it had been shown that discrete units of potential occur in the completely dark-adapted ommatidia of limulus.

Statistical analysis of these potentials has suggested that one quantum of light will evoke one such potential or quantum bump (FUORTES and YEANDLE, 1964). However, in most preparations more than one quantum bump must occur after a pulse of light to produce a propagated nerve impulse. CONE'S findings (1964) on the behavior of the early receptor potential demonstrate that a single quantum excites a receptor in the rat's eye. However, as in the case of the study of HAGINS, this conclusion was drawn from an extrapolation of measurements performed at much higher quantum rates.

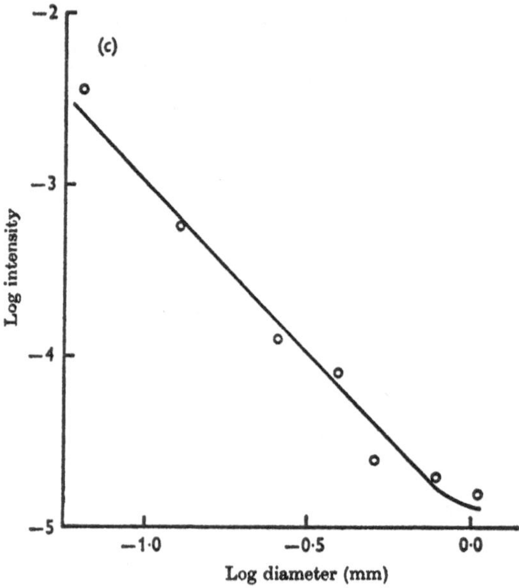

Fig. 12. The log intensity which evokes five spikes, as a function of the diameter of the disk that illuminates the receptive field's centre. The line has a slope of −2 over the central 0.8 mm. Goldfish retinal ganglion cells. (From EASTER, 1968)

A neuron that fires after a number of subliminal input pulses have summed also behaves like a coincidence scaler. In this connection the findings of KUFFLER, FITZHUGH, and BARLOW (1957) on the maintained activity in the cat's retina are suggestive. Since their spike-interval histograms were strongly skewed, they based their description on a gamma distribution, which gave a satisfactory fit. The gamma distribution they used is one that describes the output of our coincidence scalers. The lowest coincidence values or scale factors found by them were between 2 and 3.

HARTLINE'S (1957) statistical analysis of propagated nerve impulses evoked in ommatidia of limulus eye by short flashes of light indicated that the number of discrete events necessary to produce a single nerve impulse increased upon adapting the receptors to light. Moreover, this number of events or quantum bumps increased proportionately to threshold energy level for 50% frequency of occurrence of a nerve spike, thus demonstrating absence of a significant

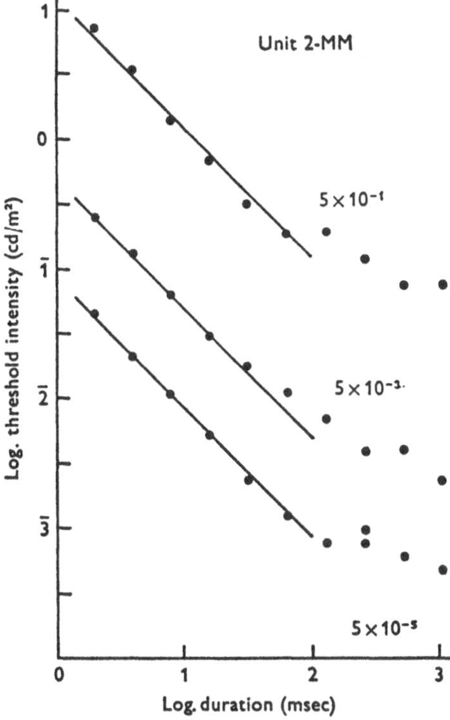

Fig. 13. Threshold intensity as a function of flash duration at several backgrounds (indicated in scotopic cd/m²). The lines on the left have slopes of -1 and therefore correspond with BLOCH's law. Beyond about 64 msec threshold intensity is proportional to a fractional power of duration, possibly $-{}^1/_2$. Cat retinal ganglions. (From LEVICK and ZACKS, 1970)

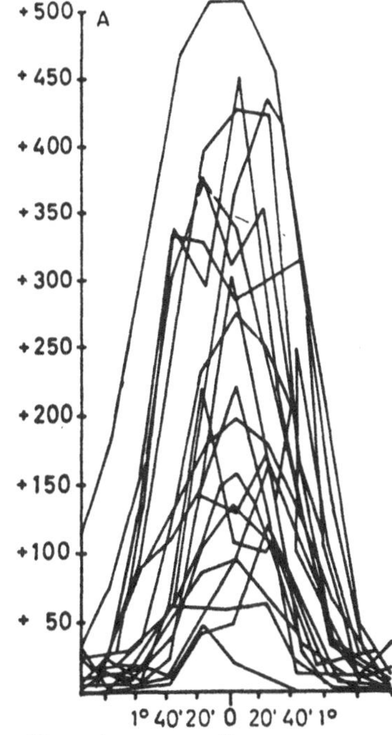

Fig. 14. Spatial response profiles of cat ganglion on-center cells; responses to a small stimulus (disc of 10′ diameter) of constant intensity shone into different parts of receptive field. (From CREUTZFELDT et al., 1970)

decrease of active photo-pigment. The lowest number of input events that could produce a single spike was found by HARTLINE to be two.

The spatial and temporal summation of the quantum coincidence concept as expressed in BLOCH'S and RICCO'S laws has frequently been the object of electrophysiological research. In most cases these laws were confirmed in cell recordings as well as in electroretinography. We refer here rather ad hoc to two recent studies: One on spatial summation of retinal ganglion cells of the goldfish retina by EASTER (1968), showing RICCO'S law up to areas 1 mm in diameter for intensities which evoked five spikes (Fig. 12); the other one on the cat's retinal ganglion cell by LEVICK and ZACKS (1970), demonstrating the validity of BLOCH'S law for luminances which made the experimenter notice a just detectable change in the maintained discharge monitored by means of a loudspeaker. This threshold effect requires a constant quantity of light energy up to flash durations of about 50 milliseconds (Fig. 13).

Probability summation, as we found it in the behavior of psychophysical thresholds, can hardly be demonstrated in the space domain by electrophysiological means. It would require simultaneous recording of the responses of a set of equivalent cells via an array of microelectrodes. However, in the time domain a single lead measuring cell responses over durations longer than BLOCH time may reveal the order of coincidence of the threshold mechanism, as it did in psychophysics. LEVICK and ZACKS (1970) found: "beyond 64 msec threshold intensity continued to fall with increasing stimulus duration such that it was now inversely proportional to a fractional power of duration (Fig. 13). It is difficult to determine the power precisely, but an inverse square root relation is not excluded" (as a twofold coincidence mechanism according to formula (1) would require).

Adapting Scaler Actions of Synaptic Junctions

Finally we mention here the spatial response profiles of the cat's retinal ganglion cells as determined by CREUTZFELDT et al. (1970). The shapes of such profiles were found to be irregular and a symmetrical bell-shaped distribution is found only occasionally. Asymmetries, plateaus around or beside the geometrical center of the receptive field center of retinal on-center ganglion cells, or pyramid like curves were found (Fig. 14). CREUTZFELDT et al. (1970) believe these irregularities to be due to a random variation in the density distributions of the dendritic trees of the cells. We suggest that these variations are not so random as suggested by these authors. An ommatidium-like organization of the retinal network can ensure that a ganglion cell is seldom centrally located amidst the group of bipolars and receptors with which it has functional connections.

Event rate reduction by an adapting scaler that follows the arithmetic strategy results in an arithmetic series of the numbers of input events needed for the successive output events. This is true for the case of a stepwise increase in average input event frequency. Increases in action potential frequency on

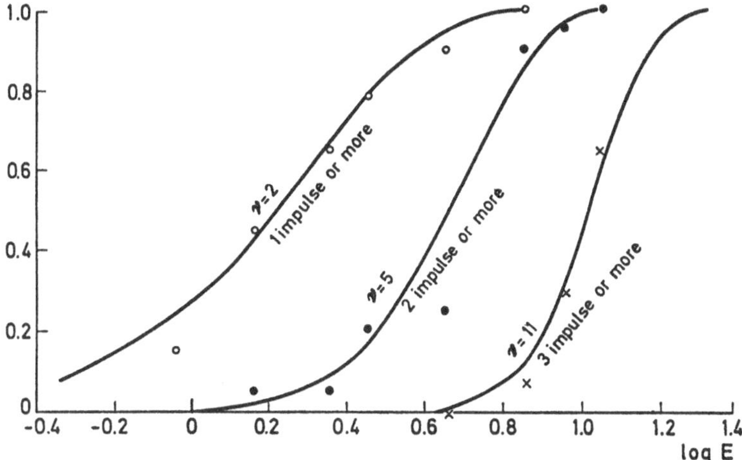

Fig. 15a. The relative frequencies of occurrence of one or more, two or more, and three or more impulses per flash as a function of the light-energy of the flashes; a: for a single receptor unit of limulus. Solid lines are theoretical curves calculated on the assumption that a test flash should contain at least the numbers of quanta indicated near the curves. (From Ratliff, 1962, from data obtained by Hartline et al., 1947)

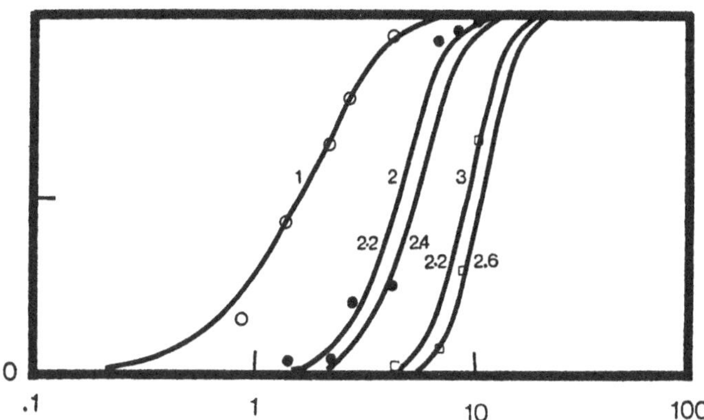

Fig. 15b. "Frequency of spike" curves for a simulated De Vries-Rose machine compared with results of Hartline et al. (1947). (From Van de Grind et al., 1970)

such a step, either upwards or downwards, are qualitatively well-known in electrophysiology as on and off effects. Mueller (1954) and Ratliff (1962) investigated the results of Hartline and coworkers (1947) on possible relationships between statistics of quantum events and the occurrence of spikes. We here give some examples of the treasures of quantitative information hidden in Hartline's beautiful work done in the early thirties, whose quality in this respect has not yet been surpassed. RATLIFF studied the relation between flash intensity and the chance of eliciting in a single nerve fiber at least one, at least two, and at least three spikes.

These relations were compared (Fig. 15a) with the probability for at least particular numbers of the Poissonean input events. In Fig. 15b we present this analysis together with simulation results obtained with a DE VRIES-ROSE device based on adapting kT scalers, as discussed in this paper. From the figure it is apparent that such a device working on the arithmetic principle reproduces HARTLINE's data quite reasonably.

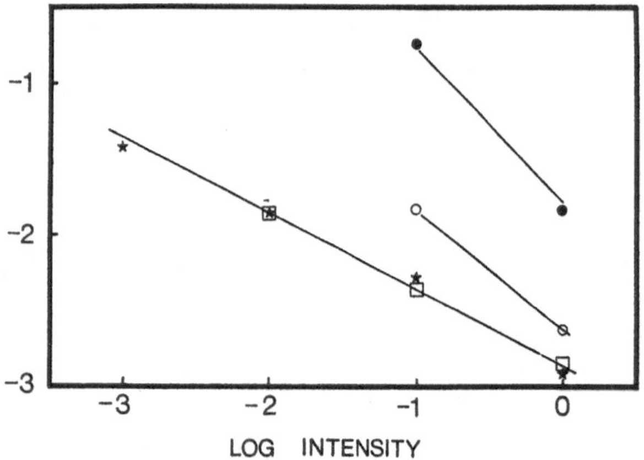

Fig. 16. Time interval τ in seconds between impulses at the beginning of a discharge as a function of stimulus intensity. Data for 0.001 (●), 0.01 (○), 0.1 (□), and 1 second (★) exposure durations from HARTLINE (1934) in MUELLER (1953). For single receptor units of limulus

MUELLER (1954), making use of HARTLINE's data, investigated the initial frequencies in bursts of spikes as a function of stimulus intensity and flash duration. As was discussed earlier, one might expect that for short flashes the number of spikes per unit time would be proportional to the input frequency. The scaling factors of the adapting scalers will not have changed in such a short time. For longer durations the scalers will adapt and we expect a different dependence between its input and output. In the DE VRIES-ROSE machines the output frequency will vary with the square root of the input frequency. In Fig. 16 we see that, for short flashes as well as for longer ones, the results agree with the behavior of our adapting scalers as based on feedback inhibition following a stage of arithmetic adaptation.

DODGE et al. (1968) studied the noise in the generator potential of visual cells in limulus, when the nerve spikes were blocked by tetrodoxein. They showed that the noise was inversely proportional to the square root of the intensity during 15-second flashes. This suggests that the effect of the quantum bumps, or the amplitude of the quantum bumps itself, depends correspondingly on intensity. This points to essentially two different possibilities towards a realization of adapting scaler machines: either the amplitude of the input events adapts, or the threshold at the synaptic junction does. These two possibilities are indistinguishable as far as the final relation between input and output frequency is concerned. Only such measurements as those of DODGE et al. (1968) can decide between them.

However, it remains doubtful whether suppression by tetrodoxein of such an essential mechanism as spike transmission leaves other properties unaffected

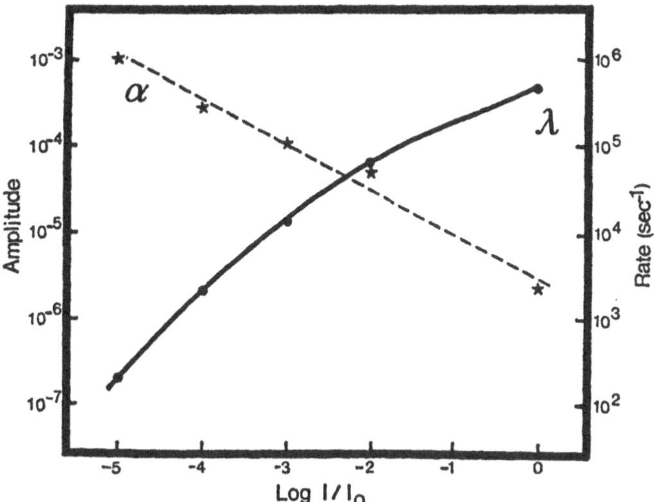

Fig. 17. Dependence of bump parameters α (amplitude) and λ (rate sec^{-1}) on light intensity I/I_0 deduced from the noise on the generator potential of limulus visual cells. The effective bump amplitude (α) is given as the fraction of the conductance of the cell in the resting state. (From DODGE et al., 1968)

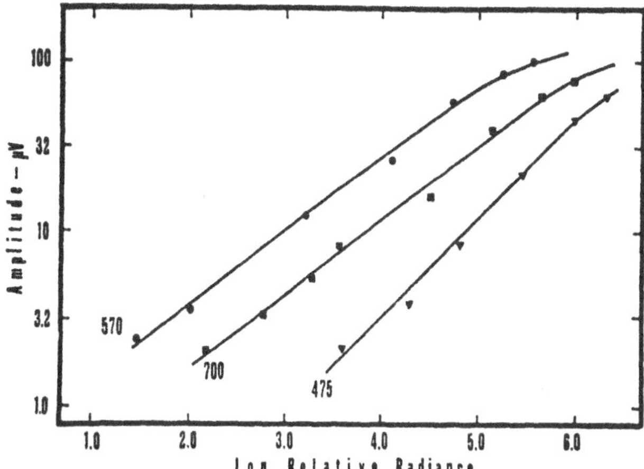

Fig. 18. Relation between log response amplitude of electroretinogram and log stimulus radiance at three wavelengths: 475, 570, 700 mµ. Flicker rate 20 Hz. The data for 475 mµ have been displaced to the right by 1.3 log units; goldfish. (From BURCKHARDT, 1968)

There is still much debate concerning where, in the visual pathways, graded responses, and where spike frequencies are the essential carriers of the visual information. We mention the possibility that in any scaler its actual momentary scale factor might very well be represented by the value of a graded potential.

This value, its variations and the spike or quantum bump frequencies are probably complementary everywhere in the network. However, it is difficult to measure both simultaneously in their whole extent for any particular junction or cell. We refer here to some recent work by MURAKAMI, SHIGEMATSU (1970) who produced some experimental evidence for the duality of conduction mechanisms in the retina.

BURCKHARDT (1968) produced evidence which indicates that the square root scalers can be found in the visual system of the goldfish. He found that the amplitude of the electroretinogram driven by flickering light of $20/\mathrm{sec}^{-1}$ was proportional to the square root of luminance level.

In the current literature there are many indications that the relationship between stimulus and response in various types of neuronal elements of the visual system is alinear. The study of CREUTZFELDT et al. (1970) that we already mentioned is a recent one.

Quantum Spike Ratios and Threshold Vision in Electrophysiology

Since the idea of a scaler adapting its scaling factor in relation to the noise in the stimulus was launched (BOUMAN, 1963), BARLOW and LEVICK (1969) have set up experiments to verify this idea by measurements of the package of quanta required for an additional spike.

I consider it quite remarkable that, with such a heavy tool as a microelectrode relative to such a subtle structure as a cat's retina, they succeeded in collecting the data presented in Fig. 19. They proceeded along the path opened by KUFFLER (1953) in the early fifties. KUFFLER, later with FITZHUGH and BARLOW (1957), made original approaches to the statistical detection of threshold signals in the retina, including rod-cone interaction. Fig. 19 demonstrates the validity of the DE VRIES-ROSE law for the production of action potentials in retinal ganglion cells: the number of quanta needed to give an extra spike is proportional to the square root of background illuminance over almost five log units. In connection with this result, BARLOW and LEVICK advocate the same type of idea as mentioned above. Their work confirms the existence of DE VRIES-ROSE machines as postulated by us. Fig. 19 shows also that in the dark-adapted retina the ganglion cell can be brought to deliver an extra potential when a package of an average of 2.7 quanta is flashed upon its on-center. Because BLOCH's law only holds for $t \leq 0.05$ seconds

(LEVICK and ZACKS, 1970), $k=1$ is excluded. It implies that two were frequently sufficient, thus representing another confirmation for a two-quantum threshold mechanism, in addition to HARTLINE's earlier one for the limulus eye, and for the probable correctness of ZWAARDEMAKER's earliest suggestion.

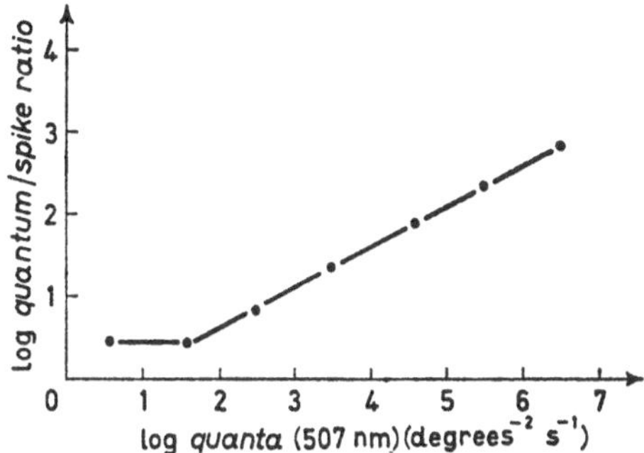

Fig. 19. Quantum-spike ratio for brief test stimuli that filled the central zone of the receptive field but did not extend onto the inhibitory surround, as a function of adapting luminance. (From BARLOW and LEVICK, 1969). Cat retinal ganglion cells

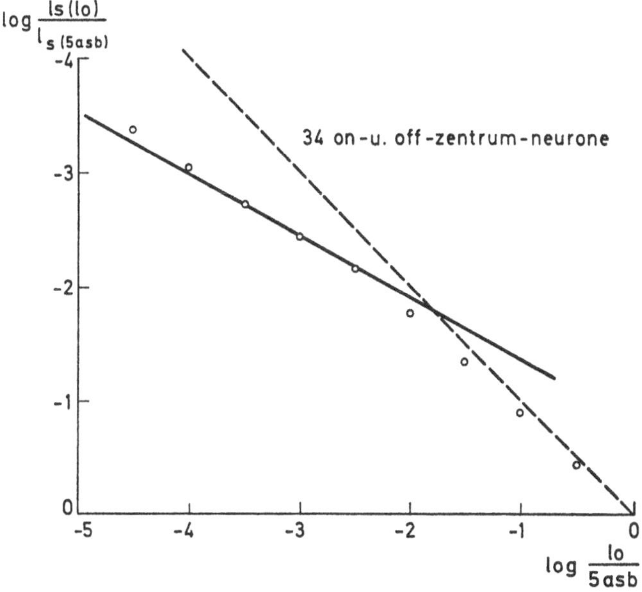

Fig. 20. Threshold intensity of cat's retinal ganglion cells as a function of adapting luminance both in relative units. Dashed curve Weber law, full curve De Vries-Rose law. Average of 34 on and off-center cells. (From FISCHER and MAY, 1970)

It also implies that there is little difference in this respect between the eye of limulus, cat, and Homo sapiens.

Recently an extensive body of information was published by FISCHER and MAY (1970), who used the same type of measurements. In their study they extended the range of luminances to the WEBER region. Their data as given in Fig. 20 show the DE VRIES-ROSE law in the first three decades of luminance and in the adjacent two illustrate WEBER behavior reasonably well.

BARLOW and LEVICK's study further indicates that the adapting scalers together with the corresponding l-detectors (KOENDERINK et al., 1971) make indeed that the output spike frequency does not depend on the magnitude of a steady input intensity.

Finally, something more may be said on rod-cone interaction in scotopic, photopic and mesopic retinal networks.

Recently ANDREWS and HAMMOND (1970a, b) have investigated cone-rod interaction in the cat's retina under different states of adaptation. In their study three main aspects can be distinguished. First, the Purkinje shift in the spectral sensitivity curve of the units concerned: the response criterion was either a just detectable variation of or a constant increase in the spike frequency of a retinal ganglion cell at the onset or offset of an increment test stimulus. Under high mesopic adaptation all responsive units show a curve that can be fitted with a pigment absorption curve 556 with varying contributions from pigment 507. At low adaptation levels a good 90% of the units receive a mixed cone-rod input; their spectral sensitivity curves can be fitted predominantly with visual pigment 507 with slight cone contamination. The remaining 10% of the units could be matched reasonably well with pigment 507 alone. Consequently, we may conclude that all units—at least, those described by these authors—show a Purkinje shift under variation of adaptation level. Therefore, at the anatomical levels they studied, there is no separation between the nerve networks in which rods and cones deposit their outputs.

The ratio between contributions of rods and cones in any particular unit was analysed by ANDREWS and HAMMOND on the basis of standard monograms of visual pigments. However, their data clearly show—and the authors confirm this—that this procedure met with serious difficulties. Conclusions on the ratio mentioned above should be considered with corresponding caution. This leaves the phenomenon of the Purkinje shift as such unaffected; it can be considered an experimental fact. At lower adaptation levels the receptors are single quantum counters. In the case of a two-quantum threshold mechanism the threshold luminance for the ommatidium group is then inversely proportional to the square of the number of its receptors. Consequently, under dark-adapted conditions the eye's spectral sensitivity curve is mainly determined by the receptor type that has the greatest density on the retinal location concerned. There also seems no objection to the general conclusion that the ratio between contributions of different types of receptors varies with adaptation level. We suggest that, for the perceptual threshold, the spatial coincidence conditions at higher adaptation level refer to single receptors so that the proportional instead of the square ratio between retinal distribution densities of different receptor types determines the spectral sensitivity curve of the eye.

The second aspect studied by ANDREWS and HAMMOND is the latency of response. Their results demonstrated that scotopic responses, when characterized by longer latencies, follow a mixed rod-cone input of 507 and 556 pigments. The shorter latency response relies predominantly on the 557 pigment receptors. However, it does not seem possible to exclude 507 pigment receptors from short latency responses. Again we refer to our conclusions from psychophysical experiments with the human eye for comparison: both rods and cones can produce scotopic as well as photopic perceptions.

Finally, the center-surround properties of short and long latency responses were studied. It was suggested that cone fields and rod fields have different retinal organizations, the networks connected with the cones extending over relatively smaller areas than those of the rods. As yet, we have not gone far into the center-surround properties of nerve networks as described by our models. Only occasionally has anything been done in this respect. It is essential to consider that in the cat's dichromatic eye the color channels may be present only at ommatidium level and not at perceptive unit level.

Thus these channels will always show more restricted spatial center-surround properties, as demonstrated in the work of ANDREWS and HAMMOND. More complete information is needed on the spectral behavior of short and long latency responses for center and surround before further conclusions can be drawn.

What the Human Eye Tells the Human Brain

In the recent literature on information processing by the visual system, the work of MATURANA et al. (1960) seemed to be a landmark. Pattern abstraction by units that have invariant capabilities for detecting special features of the stimulus have become an established idea for scientists in this field. The existence of edge and movement detectors, directional and orientational selective units and spatial frequency selectors is considered to be experimentally confirmed. So it seems hardly reasonable to ask whether this landmark is a true or a false one and whether what the frog's eye told MATURANA'S brain is right or wrong.

Nevertheless, we want to mention the possibility that the perceptive fields organized like ommatidia and the input-output relations of adapting scalers, as presented in our models, can produce pattern-abstracting properties with much less magic, although they are not specially designed for the detection of invariant parameters. It is clear that in any wiring of a distinct ommatidium-type eye "pseudo"-movement detectors and other gadgets mentioned are compulsory. Any regular two-dimensional mosaic of receptors further shows either a hexagonal or a square type of periodicity. Each receptor is a member of one or more types of functional groups. If in any retinal region elements of any particular type are mutually equivalent with respect to group participation, the periodic mosaic pattern of such functional groups must be square or hexagonal. Since their equivalence would be destroyed by any alternative random or irregular spacing. We believe this to contradict the validity of the basis for the coincidence concept for the scotopic and photopic systems.

Important consequences flow from the necessity for temporal modulations in order to maintain perception of spatial patterns and the inevitability of periodic mosaic organization in the nervous network. For instance, the transient response of a particular ganglion cell to the movement of a target may depend on the mutual relation between sizes, shapes and orientations of the target and of the retinal mosaic unit to which the ganglion cell belongs. This implies that the response depends on the direction of movement. From such dependences one may conclude that orientation-selective and direction-selective neurons exist, as LEVICK showed experimentally for the rabbit retina. The configuration dependence of scotopic spatial summation as studied by SAKITT (1971) is similarly a consequence of the ommatidium organization of the retina. For completeness, we refer to a conclusion in a previous section: contrast sensitivity is proportional to the size of the perceptive unit under test. This means that threshold is reached by stimulation of a particular number of receptors in the unit. When an edge is moved over the surface of a mosaic element, the effect of a corresponding ganglion cell might consequently be larger for a concave or convex edge than for a linear edge. This special case of mutual relation between dimension, shape and direction of movement is worth mentioning because it might explain the existence of "edge-detectors" in the frog's eye. The same possibly applies for a "bug-detector".

In OYSTER's (1968) study on preferred directions of movement of a grating in the rabbits' eye, the orthagonality of the mosaic organization can be recognized for his on-off units and the hexagonality of the organization of the on-type can also be recognized.

OYSTER (1968) also reported some results on different spatial frequencies of gratings and different speeds, indicating the possible existence of frequency-selective and speed-selective mechanisms.

Preferred frequency and speed both seem to depend on direction of movement as well. Probably the preferred values of the different stimulus aspects are interdependent. Unfortunately OYSTER reports frequency dependence and speed dependence of the unit's activity for one speed and one frequency only.

We may conclude this section with the thesis that a visual system organized in the manner of ommatidia can accommodate channels each of which is selectively sensitive to narrow bands of spatial frequencies in a very natural

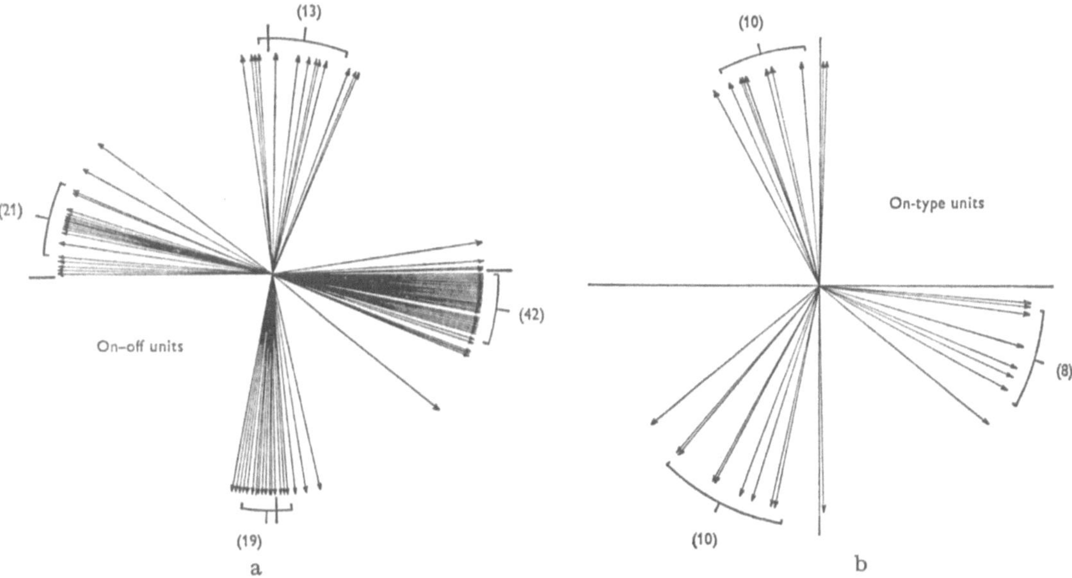

Fig. 21a. The mean preferred directions of on-off units in the visual field of the left eye of rabbits. (From OYSTER, 1968)

Fig. 21b. Distribution of preferred direction of on-type direction selective units in rabbits. (From OYSTER, 1968)

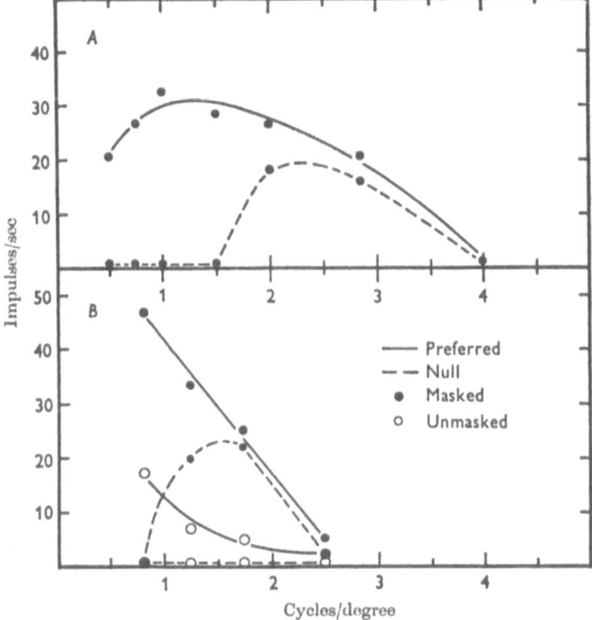

Fig. 22A and B. Gratings of different spatial frequency (cycles per degree) were moved through the receptive field, both in the preferred (continuous line) and null directions (target speed ca. 0.5 degrees/sec). A: At about 2 cpd, movement in the null direction (dashed line) begins to produce a response which rises until it equals the preferred-direction response, and then declines as the grating frequency continues to increase (top graph). B: The same experiment as in A with a different unit (filled circles). Further the response values obtained when only the receptive-field center, and not the surround was stimulated (from OYSTER, 1968) rabbit's retinal ganglion cells way. A further development of what was said about OYSTER's type of work leads to such a picture. Indeed, movement of a sinusoidal grating whose period agrees reasonably in size with the diameter of the receptive unit does not produce temporal modulation in the summed outputs of the receptors. However, if the ommatidium is covered by one or three half periods of the spatial grating, the modulation is much larger than for smaller or larger ommatidia. CAMPBELL et al. (1969), in their psychophysical as well as electrophysiological work, suppose that at any particular location in the retina a number of such small-band spatial frequency units are operating independently, subtending a total of 3 to 4 octaves. However, their data, although collected from a very large part of the retina, do not permit such a definite conclusion. This conclusion would imply that functional units of various spatial subtense are present at each retinal location. We believe this to be against the experimental evidence, for instance, FISCHER and MAY's measurements of receptive field sizes for on- as well as off-center neurons in the cat, HUBEL and WIESEL's (1960) data for the monkey and SPILLMANN's work (1964) and JUNG and SPILLMANN's study on man (1970). CAMPBELL et al. (1969) also claim to have produced evidence that all these units are multifold present at each location, in order to explain their orientational sensitivity.

In our opinion it might very well be that odd harmonic frequencies, of fundamentals belonging to a continuum of limited bandwidth, have a number of preferred locations for their detection on the retina. Such preferences would be based on the correspondence between ommatidium diameter, spatial periodicity and speed of movement of the retinal image. At each location a group of ganglion cells would serve the ommatidium, but each cell might be functionally connected to part of the receptors. This part is probably not symmetrically arranged in a circle in the unit, thus producing pseudo-directional selectivity of the cell. This could especially be so with the few red or green cones in a peripheral unit. In such a picture, directional adaptation in combination with adaptation to a spatial frequency could easily be incorporated also.

Further exploration of detection of spatial and spatio-temporal frequencies by a retinal keyboard of resonators would be intriguing, especially with respect to analogies with hearing.

Summary

Possible evidence against the rod-cone idea as a basis for the duality of the retina is discussed. It is suggested that both rods and cones may feed the chromatic or photopic, as well as the achromatic or scotopic system. Specific suggestions are made as to how a multiple coincidence ($k \geqq 2$) of singly hit receptors in a distinct group of 5–7 cones in the fovea will result in stimulation of the scotopic retinal network. A multiple ($k \geqq 2$) quantum hit in a single red or green receptor might add a chromatic or photopic red or green signal to the perceived stimulus. A more detailed analysis of color naming of monochromatic stimuli reveals that yellow signals result when the numbers of red and green signals in the group of 5–7 receptors mentioned are equal. This group of 5–7 receptors resembles the ommatidium of the insect eye. White signals might be added to yellow signals by equal numbers of red and green signals in a group of 5–7 such ommatidia. This group constitutes the perceptive unit. Each ommatidium will belong to 5–7 perceptive units because the units mutually overlap. It is suggested that blue cones feed only the photopic system. Blue signals are then elicited by single hits or by multiple quantum coincidences in the blue cones of the perceptive unit. These blue signals might inhibit yellow signals in the units so that only the white signals from the balanced red-green stimulation are left.

From subsequent experiments in the peripheral retina, corresponding suggestions are made regarding the rods. When singly hit, their signals are added to the cone signals as a contribution to the scotopic system of the ommatidium. They participate in color vision by adding white signals to the perceived stimulus. A white signal should result from a multiple hit in a single rod. At peripheral locations the ommatidium contains maximally several hundred rods and 5–7 cones.

Both the order of multiplicity of single hits required per ommatidium for a scotopic signal and those required per receptor for a photopic signal are subject to adaptational changes. In the scotopic system the multiplicity increases proportionally to the square root of the average number of quantum absorptions in the perceptive unit. In the photopic system the multiplicity is itself proportional to the average absorption per receptor. The strategy by means of which the synaptic junctions concerned arrive at these input-output relations is probably for both an inhibitory process following the "arithmetical adaptation strategy". The difference in the variation of quantum spike ratio with luminance in the two systems then results from feedback application of such an inhibitory procedure in the scotopic system and feed-forward application in the photopic system, i.e. DE VRIES-ROSE machines and WEBER machines. Most aspects of psychophysical threshold behavior are described by this model, as further analysis shows including various computer simulation studies.

In particular, quantum noise is the determinative agent for thresholds at scotopic and mesopic luminance levels and the WEBER fraction at photopic levels. A clear, mechanistic model seems to arise from this model for mesopic vision.

This picture, as a natural consequence, depicts a strict periodic mosaic as a possible essential feature at the subsequent levels of retinal information transmission.

The last part of the review discussed the results of electrophysiological studies by various other investigators, which have some bearing upon the proposed views. These references include a study of the quantum efficiency of the photo-effect in the single receptor. The discussion covers statistical phenomena in single quantum responses as they appear as quantum bumps or discrete units of potential in certain leads of the limulus eye, as well as coincidence effects in the generation of action potentials generation and spatial and temporal interaction in such coincidence effects. It is suggested that asymmetrical spatial response profiles of retinal ganglion cells are a natural consequence of ommatidium-organized eyes. Studies of input-output relations, as apparent in spike-frequencies of retinal cells, and of threshold detection in electrophysiology, show great similarity with corresponding functions in the models proposed, including the Purkinje shift of ganglion cell sensitivity under variation of adaptation. It is argued that the proposed views of a periodic ommatidium-like organization offer a rather simple frame to accommodate various "invariant" properties of the components in the visual system. More specifically, the selective properties for direction, orientation, frequency and speed are discussed. The review concludes with notes on a possible functional analogy between the retinal networks and the peripheral auditory system, up to the primary auditory nerve fibres.

References

ANDREWS, D. P., HAMMOND, P.: Mesopic increment threshold spectral sensitivity of single optic tract fibres in the cat: cone-rod interaction. J. Physiol. (Lond.) **209**, 65–83 (1970a).

ANDREWS, D. P., HAMMOND, P.: Suprathreshold spectral properties of single optic tract fibres in cat under mesopic adaptation: cone-rod interaction. J. Physiol. (Lond.) **209**, 83–103 (1970b).

BARLOW, H. B.: Retinal noise and absolute threshold. J. Opt. Soc. Amer. **46**, 634–639 (1956).

BARLOW, H. B., LEVICK, W. R.: Coding of light intensity by the cat retina. Proc. Int. School of Physics "Enrico Fermi", p. 384–396. New York: Academic Press 1969.

BAUMGARDT, E.: The quantic and statistical basis of visual excitation. J. gen. Physiol. **31**, 269–290 (1950).

BOUMAN, M. A.: Quanta explanation of vision. Docum. ophthal. (Den Haag) **4**, 23–115 (1950).

BOUMAN, M. A.: Mechanisms in peripheral darkadaptation. J. Opt. Soc. Amer. **42**, 941–950 (1952a).

BOUMAN, M. A.: Peripheral contrast thresholds for various and different wavelengths for adapting field and test stimulus. J. Opt. Soc. Amer. **42**, 820–831 (1952b).

BOUMAN, M. A.: Absolute threshold conditions for visual perception. J. Opt. Soc. Amer. **45**, 36–43 (1955).

BOUMAN, M. A.: History and present status of quantum theory in vision. In: Sensory communication. New York: Wiley & Sons 1961.

BOUMAN, M. A.: Efficiency and economy in impulse transmission in the visual system. Proc. Int. Congr. Psych. Washington (1963). Acta psychol. (Amst.) **23**, 239 (1964).

BOUMAN, M. A.: My image of the retina. Quart. Rev. Biophys. **2**, 25–64 (1969).

BOUMAN, M. A., AMPT, C. G. F.: Fluctuation theory in vision and its mechanistic model. In: Performance of the eye at low luminances. Excerpta med. Int. Congr. Series no. 125, 57–69 (1966).

Bouman, M. A., Doesschate, J. ten: The mechanism of dark adaptation. Vision Res. **1**, 386–403 (1962).

Bouman, M. A., Walraven, P. L.: A study of normal and defective color vision. N. P. L. Teddington symposium no 8, (1957a).

Bouman, M. A., Walraven, P. L.: Some color naming experiments for red and green monochromatic lights. J. Opt. Soc. Amer. **47**, 834–839 (1957b).

Bouman, M. A., Walraven, P. L.: On threshold mechanisms for achromatic and chromatic vision. In press, Acta psychologica, Amst.

Boycott, B. B., Dowling, J. E.: Organisation of the primate retina: light microscopy. Phil. Trans. B **255**, 109–184 (1969).

Brindley, G. S.: The relation of frequency of detection to intensity of stimulus for a system of many detectors each of which is stimulated by a m-quantum coincidence. J. Physiol. (Lond.) **169**, 412–415 (1963).

Burckhardt, D. A.: Cone action spectra evidence from the goldfish electroretinogram. Vision Res. **8**, 839–853 (1968).

Campbell, F. W., Cooper, G. F., Enroth-Gugell, Ch.: The spatial selectivity of the visual cells of the cat. J. Physiol. (Lond.) **203**, 223–235 (1969).

Cone, R. A.: Early receptor potential of the vertebrate retina. Nature (Lond.) **204**, 736–739 (1964).

Connors, M. M.: Luminance requirements for hue identification in small targets. J. Opt. Soc. Amer. **59**, 91–97 (1969).

Creutzfeldt, O. D., Sakmann, B., Scheich, H., Korn, H.: Sensitivity distribution and spatial summation within receptive-field center of retinal on-center ganglion cells and transfer function of the retina. J. Neurophysiol. **5**, 654–671 (1970).

Dodge, F. A., Knight, W. B., Toyoda, J.: Voltage noise in limulus visual cells. Science **160**, 88–90 (1968).

Easter, S. S.: Excitation in the goldfish retina: evidence for a non linear intensity code. J. Physiol. (Lond.) **195**, 253–281 (1968).

Fischer, B., May, H. U.: Invarianzen in der Katzenretina: Gesetzmäßige Beziehungen zwischen Empfindlichkeit, Größe und Lage receptiver Felder von Ganglienzellen. Exp. Brain Res. **11**, 448–464 (1970).

Fuortes, M. G. E., Yeandle, S.: Probability of occurrence of discrete potential waves in the eye of limulus. J. gen. Physiol. **47**, 443–464 (1964).

Graham, C. H., Yun Hsia: Saturation and the foveal achromatic interval. J. Opt. Soc. Amer. **59**, 993–997 (1969).

Grind, W. A. van de, Koenderink, J. J., Bouman, M. A.: Models of the processing of quantum signals by the human peripheral retina. Kybernetik **6**, 213–227 (1970).

Grind, W. A. van de, Koenderink, J. J., Heyde, G. L. van der, Landman, H. A. A., Bouman, M. A.: Adapting coincidence scalers and neural modelling studies of vision. Kybernetik **8**, 105–122 (1971b).

Grind, W. A. van de, Koenderink, J. J., Landman, H. A. A., Bouman, M. A.: The concepts of scaling and refractoriness in psychophysical theories of vision. Kybernetik **8**, 85–105 (1971a).

Hagins, W. A., Penn, F. D., Yoslisham, S.: Dark current and photocurrent in retinal rods. Biophys. J. 10. 380–412 (1970).

Hartline, H. K.: Intensity and duration in the excitation of single photoreceptor units. J. cell. comp. Physiol. **5**, 229 (1934).

Hartline, H. K.: Light quanta and the excitation of single receptors in the eye of limulus. Proc. Int. Congr. Photobiol. Turin, p. 103 (1957).

Hartline, H. K., Milne, L. J., Wagman, J. H.: Fluctuation of response of visual sense cell. Fed. Proc. **6**, 124 (1947).

Hecht, S., Shlear, S., Pirenne, M.: Energy, quanta and vision. J. gen. Physiol. **25**, 819–840 (1942).

Horst, G. J. C. van der, Bouman, M. A.: Spatiotemporal chromaticity discrimination. J. Opt. Soc. Amer. **59**, 1482–1488 (1969).

Hubel, D. H., Wiesel, T. N.: Receptive fields of optic nerve fibers in the spider monkey. J. Physiol. (Lond.) **154**, 572–580 (1960).

Jung, R., Spillmann, R.: Receptive field estimation and perceptual integration in human vision. In: Early experience and visual information processing in perceptual and reading disorders, p. 181–197. Washington, D.C.: Nation. Acad. Science 1970.

KOENDERINK, J. J., GRIND, W. A. VAN DE, BOUMAN, M. A.: Models of retinal signal processing at high luminances. Kybernetik **6**, 227–237 (1970).
KOENDERINK, J. J., GRIND, W. A. VAN DE, BOUMAN, M. A.: Foveal information processing at photopic luminances. Kybernetik **8**, 128–145 (1971).
KOENDERINK, J. J., GRIND, W. A. VAN DE, BOUMAN, M. A.: Opponent color coding: a mechanistic model and a new metric for color space. Kybernetik, in press.
KUFFLER, S. W.: Discharge patterns and functional organisation of mammalian retina. J. Neurophysiol. **16**, 37–68 (1953).
KUFFLER, S. W., FITZHUGH, R., BARLOW, H. B.: Maintained activity in the cat's retina in light and darkness. J. gen. Physiol. **40**, 683–702 (1957).
LEVICK, W. R., ZACKS, J. L.: Responses of cat retinal ganglion cells to brief flashes of light. J. Physiol. (Lond.) **206**, 677–700 (1970).
LIEBMAN, P. H., LEIGH, R. A.: Autofluorescence of visual receptors. Nature (Lond.) **221**, 1249–1251 (1969).
MATURANA, H. R., LETTVIN, J. Y., MCCULLOCH, W. S., PITTS, W. H.: Anatomy and physiology of vision in the frog. J. gen. Physiol. **43**, 129–176 (1960).
MCCANN, J. J., BENTON, J. L.: Interaction of the long-wave cones and the rods to produce color sensations. J. Opt. Soc. Amer. **59**, 103–107 (1969).
MUELLER, C. J.: A quantitative theory of visual excitation for the single photoreceptor. Proc. nat. Acad. Sci. (Wash.) **40**, 853–863 (1954).
MURAKAMI, M., SHIGEMATSU, Y.: Duality of conduction mechanism in bipolar cells of the frog retina. Vision Res. **10**, 1–10 (1970).
NES, F. L. VAN, KOENDERINK, J. J., NAS, H., BOUMAN, M. A.: Spatiotemporal modulation transfer in the human eye. J. Opt. Soc. Amer. **57**, 1082–1088 (1967).
OYSTER, C. W.: The analysis of image motion by the rabbit retina. J. Physiol. (Lond.) **199**, 613–635 (1968).
PIRENNE, M. H., MARRIOT, F. H. C.: Absolute threshold and frequency of seeing curves. J. Opt. Soc. Amer. **45**, 909–912 (1955).
PEDLER, C.: Duplicity theory and microstructure of the retina. In: Colour vision, Ciba Foundation Symposium, p. 52–88. London: Churchill Ltd. 1965.
PIPER, H.: Ueber die Abhängigkeit des Reizwertes leuchtender Objekte von ihrer Flächen- bzw. Winkelgröße. Z. Psychol. Physiol. Sinnesorg. **32**, 98–112 (1903).
RATLIFF, F.: Some interrelations among physics, physiology and psychology in the study of vision. In: Psychology: A study of a science, vol. 4, p. 417–482. McGraw Hill 1962.
RICCO, A.: Relazione fra il minimo angolo visuale a l'intensita luminosa. Ann. Ottal. **6**, 373–479 (1877).
ROSE, A.: The sensitivity performance of the human eye on an absolute scale. J. Opt. Soc. Amer. **38**, 196–208 (1948).
SAKITT, B.: Configuration dependence of scotopic spatial summation. J. Physiol. (Lond.) **216**, 513–529 (1971).
SCHULTZE, M.: Zur Anatomie und Physiologie der Retina. Arch. mikr. Anat. **2**, 175–286 (1866).
SPILLMANN, R.: Zur Feldorganisation der visuellen Wahrnehmung beim Menschen. Universität Münster, Ph. D. Thesis (1964).
TREZONA, P. W.: Rod participation in the "blue" mechanism and its effect on colour matching. Vision Res. **10**, 317–333 (1970).
VELDEN, H. A. VAN DER: Over het aantal lichtquanta, dat nodig is voor een lichtprikkel bij het menselijk oog. Physica **11**, 179–189 (1944).
VRIES, HL. DE: The quantum character of light and its bearing upon threshold of vision, the differential sensitivity and visual acuity of the eye. Physica **10**, 553–564 (1943).
WALD, G.: Blue-blindness in the normal fovea. J. Opt. Soc. Amer. **57**, 1289–1301 (1967).
WALRAVEN, P. L.: On the mechanisms of colour vision. Thesis, Utrecht University 1962. Institute for Perception, report SOESTERBERG, 1962.
WALRAVEN, P. L., BOUMAN, M. A.: Fluctuation theory of color discrimination of normal trichromates. Vision Res. **6**, 567–586 (1966).
WILLMER, E. N.: Duality in the retina. In: Colour vision, Ciba Foundation Symposium, p. 89–109. London: Churchill Ltd. 1965.
ZACKS, J. L.: Temporal summation phenomena at threshold and their relation to visual mechanisms. Science **170**, 197–199 (1970).
ZWAARDEMAKER, H.: Leerboek der Physiologie, deel II, 3rd edition, p. 444. Haarlem: Bohn 1921.

Namenverzeichnis

Die gewöhnlich gesetzten Ziffern weisen auf die entsprechende Stelle im Text und die *kursiven* Seitenzahlen auf das Literaturverzeichnis hin

Sachverzeichnis

Die *kursiven* Seitenzahlen beziehen sich auf Seiten, auf denen das betreffende Stichwort ausführlich behandelt wird

Ergebnisse der Physiologie

Biologischen Chemie und experimentellen Pharmakologie

Reviews of Physiology

Biochemistry and Experimental Pharmacology

Sonderdruck aus Band 65

P. Kruhøffer and Chr. Crone

Einar Lundsgaard, 1899–1968

With 1 Portrait

Nicht im Handel

Springer-Verlag Berlin · Heidelberg · New York 1972

Contents

Ergebnisse der Physiologie

Biologischen Chemie und experimentellen Pharmakologie

Reviews of Physiology

Biochemistry and Experimental Pharmacology

Sonderdruck aus Band 65

J.-S. Pitton

Mechanisms of Bacterial Resistance
to Antibiotics
With 19 Figures

Nicht im Handel

Springer-Verlag Berlin · Heidelberg · New York 1972

Contents

Ergebnisse der Physiologie

Biologischen Chemie und experimentellen Pharmakologie

Reviews of Physiology

Biochemistry and Experimental Pharmacology

Sonderdruck aus Band 65

J. B. Stanbury

Some Recent Developments in the Physiology of the Thyroid Gland

With 3 Figures

Nicht im Handel

Springer-Verlag Berlin · Heidelberg · New York 1972

Contents

Ergebnisse der Physiologie

Biologischen Chemie und experimentellen Pharmakologie

Reviews of Physiology

Biochemistry and Experimental Pharmacology

Sonderdruck aus Band 65

M. A. Bouman and J. J. Koenderink

Psychophysical Basis of Coincidence Mechanisms
in the Human Visual System
With 22 Figures

Nicht im Handel

Springer-Verlag Berlin · Heidelberg · New York 1972

Contents

Universitätsdruckerei H. Stürtz AG, Würzburg